Counterterrorism and Cybersecurity

Counterterrorism and Cybersecurity

Newton Lee

Counterterrorism and Cybersecurity

Total Information Awareness

Springer

Newton Lee
Newton Lee Laboratories, LLC
Tujunga, CA
USA

ISBN 978-1-4899-8673-3   ISBN 978-1-4614-7205-6 (eBook)
DOI 10.1007/978-1-4614-7205-6
Springer New York Heidelberg Dordrecht London

Printed on acid-free paper

Springer is part of Springer Science+Business Media (www.springer.com)

To love, peace, and freedom

About the Book

Imagine James Bond meets Sherlock Holmes: *Counterterrorism and Cybersecurity* is the sequel to *Facebook Nation* in the Total Information Awareness book series by Newton Lee. Taking no prisoners and leaving no stone unturned, the book examines U.S. counterterrorism history, technologies, and strategies from a unique and thought-provoking approach that encompasses personal experiences, investigative journalism, historical and current events, ideas from great thought leaders, and even the make-believe of Hollywood.

Demystifying Total Information Awareness, the author expounds on the U.S. intelligence community, artificial intelligence in data mining, social media and privacy, cyber attacks and prevention, causes and cures for terrorism, and longstanding issues of war and peace.

The book offers practical advice for businesses, governments, and individuals to better secure the world and protect cyberspace. It quotes U.S. Navy Admiral and NATO's Supreme Allied Commander James Stavridis: "Instead of building walls to create security, we need to build bridges". The book also provides a glimpse into the future of Plan X and Generation Z, along with an ominous prediction from security advisor Marc Goodman at TEDGlobal 2012: "If you control the code, you control the world."

Counterterrorism and Cybersecurity: Total Information Awareness will keep you up at night but at the same time give you some peace of mind knowing that "our problems are manmade—therefore they can be solved by man [or woman]," as President John F. Kennedy said at the American University commencement in June 1963.

Contents

Part I
Counterterrorism in Retrospect:
Then and Now

Chapter 1
September 11 Attacks

> We're a nation that is adjusting to a new type of war. This isn't
> a conventional war that we're waging. Ours is a campaign that
> will have to reflect the new enemy. … The new war is not only
> against the evildoers themselves; the new war is against those
> who harbor them and finance them and feed them… We will
> need patience and determination in order to succeed.
> President George W. Bush (September 11, 2001).

> Our company around the world will continue to operate in this
> sometimes violent world in which we live, offering products
> that reach to the higher and more positive side of the human
> equation.
> —Disney CEO Michael Eisner (September 11, 2001).

> He risked his life to stop a tyrant, then gave his life trying
> to help build a better Libya. The world needs more Chris
> Stevenses.
> —U.S. Secretary of State Hillary Clinton (September 12, 2012).

1.1 September 11, 2001

I was waking up in the sunny California morning on September 11, 2001. Instead
of music playing on my radio alarm clock, I was hearing fragments of news broadcast about airplanes crashing into the Pentagon and the twin towers of the World
Trade Center. I thought I was dreaming about waking up in an alternative universe.

Christopher Nolan, writer-director of *Inception* (2010), once said that "ever
since he was a youngster, he was intrigued by the way he would wake up and then,
while he fell back into a lighter sleep, hold on to the awareness that he was in fact
dreaming. Then there was the even more fascinating feeling that he could study
the place and tilt the events of the dream" [1].

N. Lee, *Counterterrorism and Cybersecurity*, DOI: 10.1007/978-1-4614-7205-6_1,   3
© Springer Science+Business Media New York 2013

Similarly, whenever I was having a nightmare, I would either semiconsciously alter the chain of events to achieve a less sinister outcome, or I would force myself to wake up and escape to reality or to a new dream state about waking up.

However, as I awoke to the radio news broadcast on the morning of September 11, I realized that it was not a lucid dream. New York City and the Pentagon were under attack. The U.S. airspace was shut down. The alternative universe was the present reality.

I went to work that morning in a state of shock and disbelief. As I entered the lobby of Disney Online, I saw a television on a portable TV cart adjacent to the reception desk. Several people were standing in front of the television. No one spoke a word. We watched the replays of an airplane crashing into the South Tower of the World Trade Center, people jumping out of the broken windows, and the collapse of the North and South Towers. It was surreal and somber.

The Disney Online office was uncharacteristically quiet that morning. I tried my best to focus on work, but my mind kept wandering off to the unfolding disasters in the East Coast and my reminiscence of the year 1984 in Virginia.

In the summer of 1984, then Virginia Tech professor Dr. Timothy Lindquist introduced me to Dr. John F. Kramer at the Institute for Defense Analyses (IDA), a nonprofit think tank serving the U.S. Department of Defense (DoD) and the Executive Office of the President [2]. A year prior, I received a surprise letter from the White House signed by President Ronald Reagan, thanking me for my support. Partially motivated by the letter, I accepted the internship offer at IDA and became a research staff member. My summer project was to assist DoD in drafting the Military Standard Common APSE (Ada Programming Support Environment) Interface Set (CAIS) [3].

My winter project was to design a counterterrorism software program for a multi-agency joint research effort involving the Defense Advanced Research Projects Agency (DARPA), National Security Agency (NSA), and Federal Bureau of Investigation (FBI). The FBI was investigating the deadly terrorist attacks against the U.S. embassy and military barrack in Beirut, Lebanon. As a co-pioneer of artificial intelligence applications in counterterrorism, I helped develop a natural language parser and machine learning program to digest news and articles in search of potential terrorist threats around the globe.

I left IDA for Bell Laboratories in 1985. However, the IDA counterterrorism project came across my mind in December 1988 when the New York-bound Pan Am Flight 103 exploded from a terrorist bomb over Lockerbie, Scotland, killing all 259 aboard and 11 on the ground [4]. The passengers included 35 Syracuse University students returning home for a Christmas break. Their surviving families were devastated by the Lockerbie bombing.

International terrorism arrived on American soil in February 1993 when a truck bomb was detonated in a public parking garage beneath the World Trade Center. "It felt like an airplane hit the building," said Bruce Pomper, a 34-year-old broker working at the Twin Towers [5]. Unbeknownst to anyone, the eyewitness' statement turned out to be an eerie prophecy. The World Trade Center bombing in 1993 and the Oklahoma City bombing in 1995 were totally eclipsed by the terrorist attacks on September 11, 2001 when the hijacked planes hit the Twin Towers and the Pentagon.

I wanted to know exactly what had happened, who was responsible, and why. I recalled one very smart computer programmer to whom Disney Online made a job offer a few years ago, but he turned it down and instead took a much higher paying job at a Wall Street firm in World Trade Center. I hoped he evacuated safely from the Twin Towers.

I checked ABCNews.com, CNN.com, and MSNBC.com, but they returned a "Server Busy" error message in the morning. The Google homepage displayed the messages: "If you are looking for news, you will find the most current information on TV or radio. Many online news services are not available, because of extremely high demand. Below are links to news sites, including cached copies as they appeared earlier today."

At 10:08 AM. Pacific Time, I received an email from Steve Wadsworth, president of Walt Disney Internet Group (WDIG):

Subject: Today
Author: Steve Wadsworth
Date: Tuesday, September 11, 2001 10:08 AM.

Team
As I'm sure you are well aware, there has been a series of tragic and frightening terrorist attacks in the United States today. I want to make sure you all know that the personal safety and security of employees is our first concern. If for any reason you have any concerns or personal issues to attend to, please do so. Coming to or staying at work is discretionary.

We are currently assessing the situation at all WDIG facilities. At this point, simply as precautionary measures:

• The 34th Street building in New York is being evacuated.
• Smith Tower in Seattle has been closed by the landlord.
• The Westin building in Seattle is only assessable with a security ID.

Should the situation change for these or any other WDIG facilities, you will be alerted.
In the meantime, please know that your safety and that of your loved ones is our top priority.

Steve

I felt like the world was getting smaller that day. Ironically, I was working on Disney's "Small World Paint" application with my colleagues Alejandro Gomez and Perigil Ilacas. But my mind kept pondering about a million things, and I ended up spending most of the day reading online news.

1.2 Disney's Responses to the 9/11 Attacks

On the morning of 9/11, Disneyland and Disney California Adventure remained closed for the entire day. The Disneyland closure marked only the second time it has locked its gates due to a national tragedy. The first time was November 23, 1963, in honor of President John F. Kennedy [6].

After the second plane hit World Trade Center on September 11, The Walt Disney Company announced the closure of its theme parks worldwide. To shut down the Disney theme parks in Florida, a "human wall procedure" was conducted at Magic Kingdom, Disney's Hollywood Studios, Animal Kingdom, and Epcot:

"Once the guests were forced to the streets of the park because all the rides were closed, all the cast members were instructed to hold hands and basically form a human wall and gently (without touching anyone) walk towards the hub of the park and eventually towards Main Street. That way we could basically force the guests out of the park. Disney Security obviously followed each human wall and made sure no one got past it" [7].

The Disney resort hotels stayed open to provide free accommodations for the stranded guests who were unable to leave since the U.S. had grounded all civilian air traffic for two days.

At 7:24 PM. Pacific Time on September 11, Disney CEO Michael Eisner sent out an email to all employees:

Subject: 09/11/01 Today's events
Author: Eisner, Michael
Date: Tuesday, September 11, 2001 7:24 PM.

Dear Fellow Cast Members:

Today's catastrophic act of violence to the World Trade Center, Pentagon, and four commercial planes is such a calamity that no words to our cast can express the sorrow or outrage that we feel. Let me offer my sincere condolences to those who lost a family member or friend or knew someone who was injured.

During this difficult time, all of our employees whose jobs require them to rise to occasions such as these in fact rose with dignity and strength. As I write this, ABC News continues to perform an outstanding job in keeping the nation informed about these deeply disturbing events, as usual with professional, informative and meticulous coverage. WABC-TV in New York, under extreme pressure including the loss of its broadcast antenna, operated with honor, as did ABC Radio.

Our parks closed for the day without incident and cast members assisted hotel guests in Florida and California, many of whom were obviously unable to get home. The Disney Stores closed as well for the day as did our Broadway shows in New York and on the road. Everybody acted with stoic determination to maintain Disney operations in efficient and caring ways.

I want to thank all of you—who are understandably upset, normally confused about our complicated world and tolerably angry—for being calm and calming to our guests. Finally let me say our company around the world will continue to operate in this sometimes violent world in which we live, offering products that reach to the higher and more positive side of the human equation.

Michael

Security tables were set up overnight at all Disney theme parks. The "crash course in flying" joke was eliminated from The Jungle Cruise ride. The Walt Disney Company Foundation established DisneyHAND: Survivor Relief Fund to provide $5 million corporate gift and $700,000 employee donations to organizations providing assistance to victims and their families. Each of the victims' families was issued a check for $1,000. Disney VoluntEARS, including trained medical personnel from Health Services and the Industrial Engineering department, helped man six of the Central Florida Blood Bank branches.

The 9/11 terrorist attacks had a big impact on all Americans and Disney employees. Ken Goldstein, General Manager of Disney Online, said at a weekly senior staff meeting, "This tragedy reminds us that the most important things in our lives are our families and friends." My colleague Eric Haseltine left his position as Executive Vice President of R&D at Walt Disney Imagineering to join the National Security Agency (NSA) as Director of Research in the following year.

1.3   FBI Warning of Hollywood as a Terrorist Target

Nine days after 9/11 on September 20, the Federal Bureau of Investigation (FBI) informed The Walt Disney Company about the possibility that a Hollywood studio could be a terrorist target. Disney president Bob Iger sent out an alarming email to all employees at 6:33 PM. Pacific time:

> Subject: 09/20/01 Important Security Notice
> Author: Iger, Bob
> Date: Thursday, September 20, 2001 6:33 PM.
>
> Dear Fellow Cast Members:
>
> Today we, along with other studios, received information from the F.B.I. that a Hollywood studio could become a target for terrorist activity.
>
> Your safety and security are our top priority and as such we are taking extraordinary precautions to ensure your safety. As you have doubtless already seen, we have already increased security at our facilities, and further steps will now be taken. We are also working diligently with the F.B.I. and local law enforcement.
>
> We urge your patience and cooperation at this time. If you do have any concerns or notice anything suspicious, please contact security at 8228-3220 (818-560-3220).
>
> Again, please know that our goal is to continue to provide a safe environment for you. Thank you again for your patience and understanding.
> Bob

Thankfully, the uncorroborated terrorist threats against Hollywood did not materialize. Nevertheless, Disney had increased security measures at all of its facilities in Burbank, Glendale, and North Hollywood. All employees were required to wear their Company I.D. while at work. Street parking adjacent to the Disney headquarters was prohibited. Visitor parking was moved to surface lots, and all incoming packages were X-rayed.

Terrorists often opt for soft targets in order to strike fear into the heart of civilians. However, it would be political suicide for a terrorist group to target children, schools, and hospitals. Nevertheless, security tables at Disney theme parks were deemed a necessity in spite of inconveniencing the visitors. America was forced to tighten security at all high-profile locations since 9/11.

On September 27, President George W. Bush became Disney's newest cheer-leader. In his speech to airline employees at O'Hare International Airport in Chicago, Bush remarked, "When they struck, they wanted to create an atmosphere of fear. And one of the great goals of this nation's war is to restore public confidence in the airline industry. It's to tell the traveling public: Get on board. Do your business around the country. Fly and enjoy America's great destination spots. Get down to Disney World in Florida. Take your families and enjoy life, the way we want it to be enjoyed" [8].

1.4 Hollywood Realism on Terrorism

Oscar Wilde wrote in his 1889 essay *The Decay of Lying* that "Life imitates Art far more than Art imitates Life" [9]. Using a commercial airliner as a weapon was not entirely a new idea. In the 1996 film *Executive Decision* starring Kurt Russell, Halle Berry, and Steven Seagal, terrorists seize control of a Boeing 747 in their plan to attack Washington DC. and cause mass causalities [10].

Premiered on November 6, 2001, the counterterrorism drama *24* set a television precedent for portraying the first African-American president (season 1–5) and the first female president (season 7 and 8). Dennis Haysbert, who played the U.S. president, predicted that his role in *24* would benefit Barack Obama in the 2008 presidential election [11]. U.S. Supreme Court Justice Clarence Thomas, Homeland Security Secretary Michael Chertoff, Defense Secretary Donald Rumsfeld, and Vice President Dick Cheney were all huge fans of *24* [12]. Because of the 9/11 attacks, the *24* premiere was postponed for one week and the footage of an airplane explosion was edited out of the show [13].

President Barack Obama is reportedly a big fan of *Homeland*, a Showtime counterterrorism television series. Iranian-born actor Navid Negahban plays terrorist Abu Nazir in *Homeland*. Negahban said, "The president is watching. It is fantastic that he is a big fan. ... But at the same time I think I will be nervous when he sends me an invitation to the White House" [14].

In fact, the U.S. intelligence community has been seeking advice from Hollywood on handling terrorist attacks [15]. In 1999, the U.S. Army established the Institute for Creative Technologies (ICT) at the University of Southern California (USC) to tap into the creative talents of Hollywood and the game industry [16].

Sometimes there can be a blurred distinction between make-believe Hollywood and real life events. In the 2011 film *God Bless America* starring Joel Murray and Tara Lynne Barr, the two lead characters shot several rude teenagers in a movie theater for talking loudly and using their cell phones during the show [17]. In July 2012, a gunman killed 12 people and injured 58 inside a movie theater in Aurora, Colorado during a midnight screening of the film *The Dark Knight Rises* [18].

Abu Jandal, al-Qaeda terrorist and former bodyguard of Osama bin Laden, was captured and held in prison before September 11, 2001. After the 9/11 attacks, the FBI interrogated Jandal to determine the identities of the hijackers and the role of bin Laden.

Jandal was shown a Yemeni news magazine with photographs of the airplanes crashing into the Twin Towers and people jumping a hundred stories off the buildings. He could hardly believe his eyes and said that they looked like a "Hollywood production" [19].

1.5 A New Day of Infamy and America at War

In the early afternoon on September 11, 2001, U.S. Senator John McCain of Arizona spoke on CNN and Fox News. McCain said, "I'm not in disagreement with others [members of Congress] who said that it is an act of war" [20]. On the night of September 11, President George W. Bush addressed the nation. Bush said, "We will make no distinction between the terrorists who committed these acts and those who harbor them. ... America and our friends and allies join with all those who want peace and security in the world; and we stand together to win the war against terrorism" [21].

On September 12, newspapers around the world headlined the 9/11 attacks. Some compared the national tragedy to the attack on Pearl Harbor and some declared America at war (see Fig. 1.1):

"New day of infamy"—*Boston Globe, Albuquerque Journal,* and *The Charleston Gazette.*
"America's Bloodiest Day: This is the second Pearl Harbor"—*Honolulu Advertiser.*
"It's War"—*New York Daily News.*
"Act of War"—*Portland Press Herald, New York Post, USA Today,* and *Sun Journal.*
"War at home"—*The Dallas Morning News.*
"War on America"—*The Daily Telegraph* (UK).
"Pearl Harbor im Jahr 2001"—*Die Welt* (Germany, Der Kommentar).

On September 23, National Security Adviser Condoleezza Rice told reporters that the U.S. government had "very good evidence of links" between Osama bin Laden's operatives "and what happened on September 11" [22]. Secretary of State Colin Powell remarked that the evidence also came from the investigation of the USS *Cole* bombing in Yemen. The Bush administration persuaded most of the world, including some Muslim nations, that a military response was justified.

On October 7, the United States and its allies began the invasion of Afghanistan under "Operation Enduring Freedom" in an attempt to capture Osama bin Laden and to force the Taliban government to hand him over to the U.S. [23].

The invasion further heightened security at high-profile American businesses nationwide and overseas. On October 8, the Disney corporate office in Hong Kong issued an email to all Disney employees in Asia:

Subject: Escalating Tensions
Author: Corporate-Broadcast AP at DISNEY-TV-HK
Date: 10/8/01 9:58 a.m.

Dear Team Asia,

By now, you know that war against terrorism has reached Afghanistan today with the retaliatory attacks by US and British forces.

It is of course expected that retaliatory attacks by the terrorists will follow across the globe. The back-and-forth has begun. In addition, as has been reported over the past month, this battle will likely go on for some time.

Fig. 1.1 Newspaper headlines on September 11, 2001

We are of course in contact with the American Consulate in Hong Kong and local officials to make sure we stay updated on any security issues or news related to us locally as this battle grows. In addition, we are in constant communication with Burbank. In all cases, there is a heightened sense of alert and security in place now.

The main thing that each of us can do is to stay alert and pay attention to the situation. As has been the plan, we will move forward with business the way we do each and every day.

Now, more than ever, it's key for each division and person within Disney to work closely together for the mutual benefit of the company. Let's help each other out whenever possible. Safety at home, in the office and during travel is our key priority while we move forward with our business—our being a family entertainment company.

Please let me know if you have any question or concerns. We'll keep you posted as we get more information.

Thank you,

Jon

On October 26, President George W. Bush signed into law the USA PATRIOT (Uniting and Strengthening America by Providing Appropriate Tools Required to Intercept and Obstruct Terrorism) Act of 2001 [24]. The Patriot Act expanded the power of law enforcement agencies to gather intelligence within the United States via surveillance of communications, financial transactions, and other activities.

On November 13, President Bush issued a military order for detention, treatment, and trial of certain non-citizens in the war against terrorism [25]. The Bush Administration established in 2002 the controversial detainment and interrogation facility at the Guantánamo Bay Naval Base in southeastern Cuba where 779 people had been detained [26].

Although Operation Enduring Freedom in Afghanistan (OEF-A) failed to locate Osama bin Laden, it toppled the Taliban government who provided sanctuary to bin Laden and his "al-Qaeda" ("The Base") network. Osama bin Laden had evaded capture for nearly 10 years before he was killed by U.S. Navy SEALs in Abbottabad, Pakistan on May 2, 2011 [27]. The war in Afghanistan continued after bin Laden's death; and it has become the longest war in American history, surpassing Iraq (8 years and 7 months) and Vietnam (8 years and 5 months) [28].

1.6   September 11, 2012

The intense hunt for terrorists over the past decade has diminished the al-Qaeda network, but the threat of terrorism is far from gone. On September 11, 2012, terrorists attacked the U.S. Consulate in Benghazi, Libya during the massive protests across the Muslim world over an anti-Islam film by Egyptian-American Mark Basseley Youssef [29]. Among the casualties were U.S. ambassador J. Christopher Stevens, ex-Navy SEALs Glen Doherty and Tyrone Woods, and an online gaming maven Sean Smith [30].

London-based think tank Quilliam believed that al-Qaeda "came to avenge the death of Abu Yahya al-Libi, al-Qaeda's second in command killed a few months ago [by a U.S. drone strike in Pakistan]" [31]. The U.S. had obtained information that Ansar al-Sharia, al-Qaeda in the Islamic Maghreb, and the Egypt based Muhammad Jamal group were likely responsible for the terrorist attack [32]. Former CIA director David Petraeus initially gave the impression that the attack arose out of a spontaneous demonstration against an anti-Islam film, but he later testified in Congress that it was an act of terrorism committed by al-Qaeda linked militants [33].

Ambassador J. Christopher Stevens, a former Peace Corps volunteer, was well-regarded among Libyans for his support of democratic transition in Libya against dictator Muammar Gaddafi's regime. "In the early days of the Libyan revolution, I asked Chris to be our envoy to the rebel opposition," said U.S. Secretary of State Hillary Clinton. "He arrived on a cargo ship in the port of Benghazi and began building our relationships with Libya's revolutionaries. … He risked his life to stop a tyrant, then gave his life trying to help build a better Libya. The world needs more Chris Stevenses" [34].

In September 2001, President George W. Bush described the long, drawn-out war on terror: "We're a nation that is adjusting to a new type of war. This isn't a conventional war that we're waging. Ours is a campaign that will have to reflect the new enemy. … These are people who strike and hide, people who know no borders … The new war is not only against the evildoers themselves; the new war is against those who harbor them and finance them and feed them. We will need patience and determination in order to succeed" [8].

References

1. **Boucher, Geoff.** 'Inception' breaks into dreams. [Online] Los Angeles Times, April 4, 2010. http://articles.latimes.com/2010/apr/04/entertainment/la-ca-inception4-2010apr04.
2. **Institute for Defense Analyses.** IDA's History and Mission. [Online] [Cited: August 29, 2012.] https://www.ida.org/aboutus/historyandmission.php.
3. **Ada Joint Program Office.** Military Standard Common APSE (Ada Programming Support Environment) Interface Set (CAIS). [Online] Defense Technical Information Center, 1985. http://books.google.com/books/about/Military_Standard_Common_APSE_Ada_Progra.html ?id=EjEYOAAACAAJ.
4. **The Guardian staff and agencies.** Lockerbie bombing—timeline. [Online] The Guardian, May 20, 2012. http://www.guardian.co.uk/uk/2012/may/20/time-line-lockerbie-bombing-megrahi.
5. **BBC.** World Trade Center bomb terrorises New York. [Online] BBC, February 26, 1993. http://news.bbc.co.uk/onthisday/hi/dates/stories/february/26/newsid_2516000/2516469.stm.
6. **Rivera, Heather Hust.** Disneyland Resort Remembers. [Online] Disney.com, November 23, 2009. http://disneyparks.disney.go.com/blog/2009/11/disneyland-resort-remembers/.
7. **Hill, Jim.** What Was It Like at Walt Disney World on 9/11. [Online] Huffington Post, September 7, 2011. http://www.huffingtonpost.com/jim-hill/what-was-it-like-at-walt-_b_952645.html.
8. **Office of the Press Secretary.** At O'Hare, President Says "Get On Board". [Online] The White House, September 27, 2001. http://georgewbush-whitehouse.archives.gov/news/releases/2001/09/20010927-1.html.
9. **Wilde, Oscar.** The Decay of Lying. [Online] The Nineteenth Century, January 1889. http://co gweb.ucla.edu/Abstracts/Wilde_1889.html.
10. **IMDb.** Executive Decision. [Online] IMDb, March 15, 1996. http://www.imdb.com/title/tt0116253/.
11. **Reynolds, Simon.** Haysbert: '24' president helped Obama. [Online] Digital Spy, July 2, 2008. http://www.digitalspy.com/celebrity/news/a106486/haysbert-24-president-helped-obama.html.
12. **Tapper, Jack.** Conservative Lovefest for '24'. [Online] ABC, June 23, 2006. http://abcnews.go.com/Nightline/story?id=2112624&page=1&singlePage=true.
13. **24 Spoilers.** Official 24 Season 1 Trailer. [Online] 24 Spoilers, January 28, 2011. http://www.24spoilers.com/2011/01/28/official-24-season-1-trailer/.
14. **Kelly, Suzanne.** Mistaken for a terrorist: Homeland star only plays one on TV. [Online] CNN, October 5, 2012. http://security.blogs.cnn.com/2012/10/05/mistaken-for-a-terrorist-homeland-star-only-plays-one-on-tv/.

15. **BBC News.** Army turns to Hollywood for advice. [Online] BBC News, October 8, 2001. http://news.bbc.co.uk/2/hi/entertainment/1586468.stm.
16. **USC ICT.** ICT Overview. [Online] University of Southern California. [Cited: December 26, 2012.] http://ict.usc.edu/about/.
17. **IMDb.** God Bless America. [Online] IMDb, May 31, 2012. http://www.imdb.com/title/tt1912398/.
18. **Frosch, Dan and Johnson, Kirk.** Gunman Kills 12 in Colorado, Reviving Gun Debate. [Online] The New York Times, July 20, 2012. http://www.nytimes.com/2012/07/21/us/shooting-at-colorado-theater-showing-batman-movie.html?pagewanted=all.
19. **Wright, Lawrence.** The Agent. Did the C.I.A. stop an F.B.I. detective from preventing 9/11? [Online] The New Yorker, July 10, 2006. http://www.newyorker.com/archive/2006/07/10/060710fa_fact_wright?currentPage=all.
20. **Fox News.** 9/11/01 Fox News Brit Hume interviews Senator John McCain. [Online] YouTube, September 11, 2001. http://www.youtube.com/watch?v=5ZCY8Rthpok.
21. **Bush, George W.** George W. Bush The Night of 9-11-01. [Online] YouTube, September 11, 2001. http://www.youtube.com/watch?v=XbqCquDl4k4.
22. **Perlez, Jane and Weiner, Tim.** U.S. to Publish Terror Evidence on bin Laden. [Online] The New York Times, September 24, 2001. http://www.nytimes.com/2001/09/24/international/24DIPL.html?pagewanted=all.
23. **CNN.** Bush announces opening of attacks. [Online] CNN, October 7, 2001. http://articles.cnn.com/2001-10-07/us/ret.attack.bush_1_qaeda-targets-al-kandahar.
24. **Bush, George W.** President Bush Signs Anti-Terrorism Bill. [Online] PBS Newshour, October 26, 2001. http://www.pbs.org/newshour/updates/terrorism/july-dec01/bush_terrorismbill.html.
25. **Office of the Press Secretary.** President Issues Military Order. [Online] The White House, November 13, 2001. http://georgewbush-whitehouse.archives.gov/news/releases/2001/11/20011113-27.html.
26. **Yasui, Hiromi.** Guantánamo Bay Naval Base (Cuba). [Online] The New York Times, November 30, 2012. http://topics.nytimes.com/top/news/national/usstatesterritoriesandpossessions/guantanamobaynavalbasecuba/index.html.
27. **Phillips, Macon.** Osama Bin Laden Dead. [Online] The White House Blog, May 2, 2011. http://www.whitehouse.gov/blog/2011/05/02/osama-bin-laden-dead.
28. **Dermody, William.** The Longest War. [Online] USA Today. http://www.usatoday.com/news/afghanistan-ten-years-of-war/index.html.
29. **Reuters.** Terrorists killed U.S. ambassador to Libya: Panetta. [Online] Chicago Tribune, September 27, 2012. http://www.chicagotribune.com/news/sns-rt-us-libya-usa-investigation-bre88q1jw-20120927,0,3904573.story.
30. **Smith, Matt.** Ex-SEALs, online gaming maven among Benghazi dead. [Online] CNN, September 15, 2012. http://www.cnn.com/2012/09/14/us/benghazi-victims/index.html.
31. **Quilliam.** The Attack on the US Consulate Was A Planned Terrorist Assault Against US and Libyan Interests. [Online] Quilliam, September 12, 2012. http://www.quilliamfoundation.org/press-releases/the-attack-on-the-us-consulate-was-a-planned-terrorist-assault-against-us-and-libyan-interests/.
32. **Hosenball, Mark.** Congress to continue probes of Benghazi attacks. [Online] Chicago Tribune, November 7, 2012. http://www.chicagotribune.com/news/politics/sns-rt-us-usa-campaign-benghazibre8a62cg-20121107,0,1506034.story.
33. **CNN Wire Staff.** Ex-CIA chief Petraeus testifies Benghazi attack was al Qaeda-linked terrorism. [Online] CNN, November 16, 2012. http://www.cnn.com/2012/11/16/politics/benghazi-hearings/index.html.
34. **Pearson, Michael.** Slain ambassador died 'trying to help build a better Libya'. [Online] CNN, September 15, 2012. http://www.cnn.com/2012/09/12/world/africa/libya-us-ambassador-killed-profile/index.html.

Chapter 2
U.S. Intelligence Community

Secrecy stifles oversight, accountability, and information sharing.
—The 9/11 Commission (July 22, 2004).

In solving intelligence problems, including diversity of thought is essential.
—Letitia "Tish" Long, Director of the National Geospatial-Intelligence Agency (2012).

The situation was, and remains, too risky to allow someone to experiment with amateurish, Hollywood style interrogation methods—that in reality—taints sources, risks outcomes, ignores the end game, and diminishes our moral high ground in a battle that is impossible to win without first capturing the hearts and minds around the world. It was one of the worst and most harmful decisions made in our efforts against al-Qaeda.
—Former FBI Agent Ali Soufan (May 13, 2009).

2.1 FBI Ten Most Wanted Fugitive: Osama Bin Laden

In August 1996, *Al Quds Al Arabi* in London published Osama bin Laden's fatwā entitled "Declaration of War against the Americans Occupying the Land of the Two Holy Places" [1]. The land referred to Saudi Arabia and the two holy places Mecca and Medina. In February 1998, bin Laden issued a second fatwā declaring a holy war or jihad against America and Israel [2]. In August 1998, al-Qaeda bombed the U.S. embassies in Kenya and Tanzania simultaneously, killing 224 people [3]. In December 1998, Director of Central Intelligence's (DCI) Counterterrorist Center (CTC) reported to President Bill Clinton that "Bin Ladin and his allies are preparing for attacks in the US, including an aircraft hijacking" [4].

N. Lee, *Counterterrorism and Cybersecurity*, DOI: 10.1007/978-1-4614-7205-6_2,   15
© Springer Science+Business Media New York 2013

In June 1999, two years before 9/11, Osama bin Laden (aka Usama Bin Laden) was placed on the FBI's Ten Most Wanted Fugitives list [5]. (See Fig. 2.1). Bin Laden was wanted for the murder of U.S. nationals outside the U.S., in connection with the August 1998 bombings of the U.S. Embassies.

After the September 11 attacks, the U.S. Department of State offered a reward of up to $25 million for information leading directly to the apprehension or conviction of Osama bin Laden. An additional $2 million was offered through a

FBI TEN MOST WANTED FUGITIVE

Murder of U.S. Nationals Outside the United States; Conspiracy to Murder U.S.
Nationals Outside the United States; Attack on a Federal Facility Resulting in Death

USAMA BIN LADEN

Deceased

Date of Photograph
Unknown

Aliases:
Usama Bin Muhammad Bin Ladin, Shaykh Usama Bin Ladin, the Prince, the Emir, Abu Abdallah, Mujahid Shaykh, Hajj, the Director

DESCRIPTION

Date(s) of Birth Used:	1957	Hair:	Brown
Place of Birth:	Saudi Arabia	Eyes:	Brown
Height:	6'4" to 6'6"	Complexion:	Olive
Weight:	Approximately 160 pounds	Sex:	Male
Build:	Thin	Nationality:	Saudi Arabian
Occupation:	Unknown		
Scars and Marks:	None known		
Remarks:	Bin Laden is the leader of a terrorist organization known as Al-Qaeda, "The Base". He is left-handed and walks with a cane.		

CAUTION

Usama Bin Laden is wanted in connection with the August 7, 1998, bombings of the United States Embassies in Dar es Salaam, Tanzania, and Nairobi, Kenya. These attacks killed over 200 people. In addition, Bin Laden is a suspect in other terrorist attacks throughout the world.

REWARD

The Rewards For Justice Program, United States Department of State, is offering a reward of up to $25 million for information leading directly to the apprehension or conviction of Usama Bin Laden. An additional $2 million is being offered through a program developed and funded by the Airline Pilots Association and the Air Transport Association.

CONSIDERED ARMED AND EXTREMELY DANGEROUS

If you have any information concerning this person, please contact your local FBI office or the nearest American Embassy or Consulate.

June 1999 Poster Revised November 2001

Fig. 2.1 FBI ten most wanted fugitive—Usama Bin Laden. (Courtesy of the Federal Bureau of Investigation)

program developed and funded by the Airline Pilots Association and the Air Transport Association. However, the revised FBI poster in November 2001 inexplicably did not mention 9/11 in bin Laden's list of crimes. Instead, it simply stated, "In addition, Bin Laden is a suspect in other terrorist attacks throughout the world."

Bin Laden initially denied any involvement in 9/11. In a statement issued to the Arabic satellite channel Al Jazeera, bin Laden said, "The U.S. government has consistently blamed me for being behind every occasion its enemies attack it. I would like to assure the world that I did not plan the recent attacks, which seems to have been planned by people for personal reasons. I have been living in the Islamic emirate of Afghanistan and following its leaders' rules. The current leader does not allow me to exercise such operations" [6].

To contradict bin Laden's denial, the Bush administration released in December 2001 a video tape showing bin Laden bragging about the 9/11 attacks: "I was thinking that the fire from the gas in the plane would melt the iron structure of the building and collapse the area where the plane hit and all the floors above it only. This is all that we had hoped for" [7].

Nonetheless, Saudi Defense Minister Prince Sultan bin Abdul Aziz questioned, "Who stands behind this terrorism and who carried out this complicated and carefully planned terrorist operation? Are bin Laden and his supporters the only ones behind what happened or is there another power with advanced technical expertise that acted with them?" [8].

In an audiotape surfaced in May 2006, bin Laden claimed responsibility for 9/11: "The truth is that he [Zacarias Moussaoui] has no connection whatsoever with the events of September 11th, and I am certain of what I say, because I was responsible for entrusting the 19 brothers—Allah have mercy upon them—with those raids, and I did not assign brother Zacarias to be with them on that mission" [9].

The initial denials and subsequent admissions might have suggested that bin Laden eventually decided to go for this once-in-a-lifetime opportunity and take credit for the actions of his top lieutenants in order to maintain a unified support of jihad against America. Although the authenticity of the bin Laden "confession" audio and video tapes is subject to debate, the al-Qaeda terrorist organization has been inextricably linked to the 9/11 attacks, thanks to the FBI's relentless investigations.

2.2 An American Hero Born in Lebanon

Back in March 2000, the U.S. Central Intelligence Agency (CIA) learned that two military trained al-Qaeda terrorists, Khalid al-Mihdhar and Nawaf al-Hazmi, had entered the United States [10]. However, the CIA did not inform the FBI or the U.S. State Department until August 2001, a year and half later [11]. By then, it was too late for the FBI to track down al-Mihdhar and al-Hazmi who, along with three other terrorists, hijacked American Airlines Flight 77 that crashed into the Pentagon on September 11, killing 189 people [12].

"When [Ali] Soufan realized that the CIA had known for more than a year and a half that two of the hijackers were in the country he ran into the bathroom and threw up," wrote Pulitzer Prize-winning author Lawrence Wright in the July 2006 issue of *The New Yorker* [13].

Lebanese-American Ali Soufan is a former FBI agent who might have prevented 9/11 if the CIA and the FBI had been cooperative with one another. Soufan's close friend and colleague John O'Neill was killed in the attacks. O'Neill was the former head of the FBI National Security Division. He died less than a month after he started his new job as head of security at the World Trade Center.

Back in 1998, O'Neill drafted Soufan into the FBI's I-49 squad to investigate al-Qaeda in connection with the U.S. embassy bombings in Kenya and Tanzania [3]. Soufan was one of eight FBI agents who spoke Arabic fluently. The FBI collected evidence linking the bombings to Osama bin Laden, and placed him on the FBI's Ten Most Wanted Fugitives list [5]. The FBI also discovered a phone number in Yemen that functioned as a virtual switchboard for al-Qaeda's global organization. However, the CIA had jurisdiction over monitoring the overseas phone conversations, and they refused to share the intercepted messages with the FBI.

The CIA followed al-Qaeda terrorist Khalid al-Mihdhar to Dubai where they learned that he had a U.S. visa. Nevertheless, the CIA did not alert the FBI or the U.S. State Department to put al-Mihdhar on the terrorist watch list. The CIA had hoped to recruit al-Mihdhar to infiltrate the al-Qaeda organization. However, the CIA realized in March 2000 that al-Mihdhar and his accomplice Nawaf al-Hazmi had entered the U.S. legally. Nonetheless, the CIA failed to inform the FBI and the U.S. State Department that had jurisdiction inside the United States.

In October 2000, a fiberglass fishing boat containing plastic explosives blasted open a hole in USS *Cole* in Yemen, killing 17 American sailors [14]. O'Neill and Soufan went to Yemen for the investigation. The FBI identified the one-legged al-Qaeda lieutenant Khallad who orchestrated the *Cole* attack. Despite numerous requests from the FBI, the CIA withheld information about Khallad and his meeting in Malaysia with the future 9/11 hijackers.

O'Neill retired from the FBI in August 2001 and took a job as head of security at the World Trade Center. Meanwhile, the FBI learned that al-Mihdhar and al-Hazmi were in the U.S. but the agency was unable to track them down.

Less than a month later on September 11, al-Mihdhar and al-Hazmi hijacked American Airlines Flight 77 that crashed into the Pentagon, killing a total of 189 people [12]. Other 9/11 hijackers piloted American Airlines Flight 11 and United Airlines Flight 175 that crashed into the North Tower and South Tower of the World Trade Center, killing O'Neill and a total of 2,762 people [15]. Including the crash of United Airlines Flight 93 after its 40 passengers revolted against the four hijackers [16], the total death toll on September 11, 2001 was 2,995 people.

On September 12, the FBI ordered its agents to identify the 9/11 hijackers "by any means necessary." Soufan was tasked to interrogate al-Qaeda terrorist Fahd al-Quso in Yemen, the only lead that the FBI had at the time. The CIA handed Soufan the surveillance photos of 20 al-Qaeda terrorists and a complete report on the Malaysia meeting of the suspected hijackers. When Soufan learned that the CIA had known for more than a year and a half that two al-Qaeda terrorists were living in the U.S., the shock made him physically ill.

Unlike Jack Bauer (played by Kiefer Sutherland) in the counterterrorism TV drama 24, Soufan did not torture or threaten the suspects in order to extract information from them and exact revenge for the death of his close friend and colleague O'Neill. Instead, Soufan's arsenal of interrogation included American bottle water, sugarless wafers, theological debates, and a history book of America in the Arabic language.

Soufan was able to persuade Fahd al-Quso and disarm Abu Jandal (aka Nasser al-Bahri) who was trained in counter-interrogation techniques. Al-Quso and Jandal eventually identified many 9/11 hijackers from the photos of known al-Qaeda terrorists, including Mohammed Atta who was the lead hijacker. Working on the 9/11 case for days and nights with almost no sleep, his coworkers referred to Soufan as "an American hero."

For every hero recognized by the news media, there are many more unsung heroes who work steadfastly and risk their lives to keep the world safer. In 2009, for example, seven CIA officers were killed by a Jordanian double agent when he detonated his suicide bomb vest at their secret meeting in Khost, Afghanistan [17].

2.3 The FBI-CIA Bureaucratic Rivalries

Former *Newsweek* investigative correspondent Michael Isikoff said, "The CIA knew who they were, they knew that they were suspected al-Qaeda operatives, they failed to alert the INS [Immigration and Naturalization Service], the State Department, the Customs Service, agencies who could have kept them out of the country. And, perhaps more importantly, they failed to alert the FBI, which could have tracked them while they were in the country" [10].

Although the Ali Soufan story placed the blame squarely on the CIA, Michael Scheuer, former chief of the Osama bin Laden unit at the CIA Counterterrorism Center, accused the FBI instead. Scheuer wrote in the July 4, 2006 issue of *The Washington Times* [18]:

> FBI officers sat in the unit I first commanded and then served in and they read the same information I did. If the data did not get to FBI headquarters it is because the FBI then lacked, and still lacks, a useable computer system. The FBI did not know the September 11 hijackers were here because Judge Louis Freeh [5th FBI Director] and Robert Mueller [6th FBI Director] have failed to provide their officers computers that allow them to talk securely to their headquarters and other intelligence community elements. ….. In my own experience, Mr. O'Neill was interested only in furthering his career and disguising the rank incompetence of senior FBI leaders. He once told me that he and the FBI would oppose an operation to capture bin Laden and take him to a third country for incarceration. When I asked why, he replied, Why should the FBI help to capture bin Laden if the bureau won't get credit among Americans for his arrest and conviction?

Contrary to Scheuer's negative assessment of O'Neill who was killed in the 9/11 attacks, the FBI I-49 squad intercepted phone calls between bin Laden and his al-Qaeda operatives by installing two antennae in the Pacific and the Indian Ocean, as well as a satellite phone booth with a hidden camera in Kandahar, Afghanistan [13].

The CIA apparently did not get along with the National Security Agency (NSA) either. A military officer from the 1998 Middle East signals-intelligence operation team told Seymour Hersh of *The New Yorker* that he was unable to discuss the activity with representatives of the CIA and the NSA at the same time. "I used to meet with one in a safe house in Virginia, break for lunch, and then meet with the other," the officer said. "They wouldn't be in the same room" [19].

A senior manager at a U.S. intelligence agency disclosed that the major intelligence centers were so consumed by internecine warfare that the professional analysts find it difficult to do their jobs. "They're all fighting among each other," said the senior manager. "There's no concentration on issues" [19].

To make matter worse, the U.S. Justice Department in 1995 established a policy known as "the Wall" or "Chinese Wall," that hindered the exchange of foreign intelligence information between FBI agents and criminal investigators. U.S. Attorney Mary Jo White voiced her dissonance in a memorandum faxed to Deputy Attorney General Jamie Gorelick in December 1995. White argued that the Justice Department and the FBI were structured and operating in a way that did not make maximum legitimate use of all law enforcement and intelligence avenues to prevent terrorism and prosecute terrorists. White asserted that the Justice Department was building "unnecessary and counterproductive walls that inhibit rather than promote our ultimate objectives" and that "we must face the reality that the way we are proceeding now is inherently and in actuality very dangerous" [20]. Using the Chinese Wall policy as an excuse, the FBI withheld intelligence from the White House, and the CIA refused to share intelligence with the FBI.

In his 1998 book *Secrecy: The American Experience*, U.S. Senator Daniel Patrick Moynihan wrote that "Departments and agencies hoard information, and the government becomes a kind of market. Secrets become organizational assets, never to be shared save in exchange for another organization's assets. ... Too much of the information was secret, not sufficiently open to critique by persons outside government. Within the confines of the intelligence community, too great attention was paid to hoarding information, defending boundaries, securing budgets, and other matters of corporate survival. ... The system costs can be enormous. In the void created by absent or withheld information, decisions are either made poorly or not at all" [21].

2.4 Operational Failures of the U.S. Intelligence Community

After 9/11, CIA director George Tenet and CIA Counterterrorist Center director Cofer Black testified to Congress that the CIA had divulged information about future hijacker Khalid al-Mihdhar's to the FBI in a timely manner. However, the 9/11 Commission concluded that their statements were false [22].

The 9/11 Commission Report identified several major operational failures—opportunities that the U.S. intelligence community and governmental agencies could have exploited to prevent 9/11 [23]. The failures included:

Central Intelligence Agency (CIA)

- Not watchlisting future hijackers Khalid al-Mihdhar and Nawaf al-Hazmi, not trailing them after they traveled to Bangkok, and not informing the FBI about one future hijacker's U.S. visa or his companion's travel to the United States.
- Not sharing information linking individuals in the *Cole* attack to al-Mihdhar.

Federal Bureau of Investigation (FBI)

- Not taking adequate steps in time to find al-Mihdhar or al-Hazmi in the United States.
- Not linking the arrest of Zacarias Moussaoui, described as interested in flight training for the purpose of using an airplane in a terrorist act, to the heightened indications of attack.
- Not expanding no-fly lists to include names from terrorist watchlists.

U.S. State Department and Immigration and Naturalization Service (INS)

- Not discovering false statements on visa applications.
- Not recognizing passports manipulated in a fraudulent manner.
- Not realizing false statements to border officials to gain entry into the United States.

Federal Aviation Administration (FAA)

- Not making use of the U.S. TIPOFF terrorist watchlist where two of the hijackers were listed.
- Not searching airline passengers identified by the Computer-Assisted Passenger Prescreening System (CAPPS).
- Not hardening aircraft cockpit doors or taking other measures to prepare for the possibility of suicide hijackings.

Furthermore, a 2001 Congressional report disclosed that the NSA was faced with "profound needle-in-the-haystack challenges." *The New York Times* revealed in 2002 that there were 200 million pieces of intelligence in a regular workday, and less than one percent of it was ever decoded, translated, or processed [24].

Despite the establishment of the Terrorist Screening Center (TSC) in 2003, U.S. agencies handling the terrorist watch lists have continued to "work from at least 12 different, sometimes incompatible, often uncoordinated and technologically archaic databases" [25]. The TSC has been tasked to consolidate the State Department's TIPOFF, Homeland Security's No-fly list, and FBI's Interpol terrorism watch list [26].

For spy agencies like the CIA and military intelligence organizations, *The Wall Street Journal* revealed in 2009 that there are hundreds of databases each and most of them are not linked up [27]. Analysts often have to query individual databases separately and analyze the data through the conventional "pen and paper" method.

In short, finger-pointing, interagency rivalries, personal animosity, operational inefficiency, outmoded government regulations, and intentional withholding of vital information had resulted in a colossal failure of the U.S. intelligence community to protect and serve the American public.

2.5 Unity of Counterterrorism Effort Across U.S. Government

Most U.S. military officers serving on the front lines overseas have heard "One Team, One Fight"—everyone working together to accomplish the mission [28]. Aesop's fable *The Four Oxen and the Lion* reminds us that "united we stand, divided we fall." To effectively fight against terrorism, the 9/11 Commission called for unity of effort in five areas of the U.S. government [23]:

1. **National Counterterrorism Center (NCTC)** unifying strategic intelligence and operational planning against Islamist terrorists across the foreign-domestic divide. Placed in the Executive Office of the President, the NCTC would:
   a. Build on the existing Terrorist Threat Integration Center, and would replace it and other terrorism "fusion centers" within the government.
   b. Become the authoritative knowledge bank with information collected from both inside and outside the United States.
   c. Perform joint operational planning with existing agencies and track implementation of plans.
   d. Influence the leadership and the budgets of the counterterrorism operating arms of the CIA, the FBI, the departments of Defense, and Homeland Security.
   e. Follow the policy direction of the President and the National Security Council.

2. **National Intelligence Director (NID)** unifying the U.S. intelligence community. Located in the Executive Office of the President, the NID would:
   a. Oversee national intelligence centers (e.g. CIA, DIA [Defense Intelligence Agency], NSA, and FBI) that combine experts from all the collection disciplines against common targets such as counterterrorism or nuclear proliferation.
   b. Oversee the agencies that contribute to the national intelligence program including setting common standards for personnel and information technology.
   c. Have authority over three intelligence deputies:
   - Director of the CIA (for foreign intelligence)
   - Under Secretary of Defense for Intelligence (for defense intelligence)
   - Executive Assistant Director for Intelligence at the FBI or Under Secretary of Homeland Security for Information Analysis and Infrastructure Protection (for homeland intelligence)

3. **Network-based Information Sharing System** unifying the many participants in the counterterrorism efforts and their knowledge. Transcending traditional governmental boundaries, the decentralized network-based information system would replace the system of "need to know" by a system of "need to share."

The 9/11 Commission acknowledged that "secrecy stifles oversight, account-ability, and information sharing." Legal and policy issues must be resolved in order for the new information sharing system to be used effectively.

4. **U.S. Congress** unifying and strengthening congressional oversight to improve quality and accountability. For intelligence oversight, Congress would combine authorizing and appropriating intelligence committees to empower their over-sight function. To minimize national security risks during transitions between administrations, Congress would create a permanent standing committee for homeland security in each chamber.

5. **U.S. Government** strengthening the FBI and homeland defenders. The FBI would establish a specialized and integrated national security workforce consist-ing of agents, analysts, linguists, and surveillance specialists, in order to develop a deep expertise in intelligence and national security. The Department of Defense's Northern Command would regularly access the adequacy of strategies and planning to defend against military threats to the homeland. The Department of Homeland Security would regularly assess the types of threats the country faces, in order to determine the readiness of the U.S. government to respond to those threats.

Figure 2.2 shows the post-9/11 organizational chart for the U.S. intelligence community recommended by the 911 Commission to achieve a unity of effort in managing intelligence. The Executive Office of the President includes the President of the United States (POTUS), the National Intelligence Director (NID), and the National Counterterrorism Center (NCTC). The three NID deputies are the CIA Director (foreign intelligence), the Under Secretary of Defense for Intelligence (defense intelligence), and the FBI Intelligence Director (homeland intelligence).

The "new" Open Source Agency in Fig. 2.2 became a reality in 2005 when Douglas Naquin was appointed the director of the Open Source Center (OSC). Replacing the former Foreign Broadcast Information Service (FBIS) head by Naquin, the OSC collects information available from "the Internet, databases, press, radio, television, video, geospatial data, photos and commercial imagery" [29]. According to the OSC website, "OpenSource.gov provides information on foreign political, military, economic, and technical issues beyond the usual media from an ever expanding universe of open sources. Our website contains sources from more than 160 countries, in more than 80 languages and hosts content from several commercial providers, as well as content from OSC partners" [30].

2.6 Transition from Need-to-Know to Need-to-Share and Need-to-Provide

Based on the 9/11 Commission's recommendations, Congress passed the Intelligence Reform and Terrorism Prevention Act in 2004 and established the Office of the Director of National Intelligence (ODNI) [31]. Former Ambassador John Negroponte served as the first Director of National Intelligence (DNI) in April 2005.

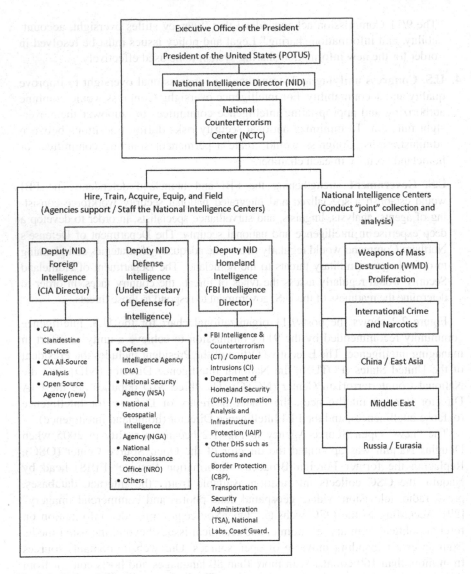

Fig. 2.2 Unity of effort in managing intelligence

Although the CIA had objected to the creation of the ODNI, the CIA issued a statement in 2006 admitting the necessity of interagency cooperation: "Based on rigorous internal and external reviews of its shortcomings and successes before and after 9/11, the CIA has improved its processing and sharing of intelligence. CIA's focus is on learning and even closer cooperation with partners inside and outside government, not on public finger pointing, which does not serve the American people well" [32].

Former NSA director and vice admiral John Michael McConnell became the second Director of National Intelligence in February 2007. He implemented an aggressive "100 Day Plan for Integration and Collaboration" in order to address and overcome barriers—legal, policy, technology, process, and culture—in the U.S. intelligence community [33]. The 100 Day Plan focused on six enterprise integration priorities:

1. Create a Culture of Collaboration.
2. Foster Collection and Analytic Transformation.
3. Build Acquisition Excellence and Technology Leadership.
4. Modernize Business Practices.
5. Accelerate Information Sharing.
6. Clarify and Align DNI's Authorities.

McConnell's 100 Day Plan was followed by his "500 Day Plan for Integration and Collaboration" in October 2007 that "continues to build the foundation to enable the IC to work as a single, integrated enterprise" [34].

As of September 2012, the U.S. intelligence community is a coalition of 17 agencies and organizations headed by the Office of the Director of National Intelligence (ODNI) [35] (See Fig. 2.3):

1. ODNI.
2. Air Force Intelligence, Surveillance, and Reconnaissance (AF ISR).
3. Army Intelligence (G-2).
4. Central Intelligence Agency (CIA).
5. Coast Guard Intelligence.
6. Defense Intelligence Agency (DIA).
7. Department of Energy (DOE).
8. Department of Homeland Security (DHS).
9. Department of State: Bureau of Intelligence and Research (INR).
10. Department of the Treasury: Office of Intelligence and Analysis (OIA).
11. Drug Enforcement Administration (DEA): Office of National Security Intelligence (ONSI)
12. Federal Bureau of Investigation (FBI): National Security Branch (NSB).
13. Marine Corps Intelligence.
14. National Geospatial-Intelligence Agency (NGA).
15. National Reconnaissance Office (NRO).
16. National Security Agency (NSA)/Central Security Service (CSS).
17. Navy Intelligence (ONI).

The intelligence community employs more than 100,000 people including private contractors. The annual budget exceeds $50 billion, more than the U.S. government spends on energy, scientific research, and the federal court and prison systems [36].

Letitia "Tish" Long, Director of the National Geospatial-Intelligence Agency (NGA), described the shift across the post-9/11 intelligence community as the transition from a need-to-know atmosphere to a need-to-share and need-to-provide culture. "In solving intelligence problems, including diversity of thought is essential," said Long in a 2012 interview. "[In] the Osama bin Laden operation, the intelligence community witnessed the true value of merging many thoughts and

Fig. 2.3 U.S. intelligence
community. (Courtesy of
the Office of the Director of
National Intelligence)

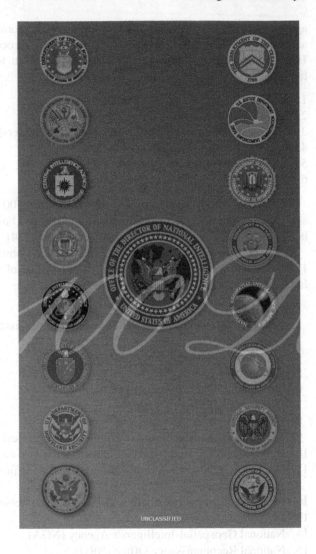

perspectives, and we must continue to replicate this kind of integration across the
enterprise in the future" [37]. The NGA was responsible for the analysis of satellite imagery of the bin Laden compound in Pakistan [38].

2.7 Chinese Wall between the CIA and FBI

At the 2012 Aspen Institute Security Conference, Admiral William McRaven
touted the bin Laden raid as one of the "great intelligence operations in history"
[39]. However, the FBI and CIA have been at odds with each other's interrogation

techniques. FBI agent Ali Soufan who firmly believes in "knowledge and empa-thy" has accused CIA's "enhanced interrogation" methods for being harsh, inef-fective, and borderline torture. Senate Intelligence Committee Chairman Dianne Feinstein and Senate Armed Services Committee Chairman Carl Levin released their findings in April 2012 that seriously questioned the effectiveness of CIA's coercive interrogation methods [40]. Ali Soufan testified before Congress in May 2009 [41]:

> The Informed Interrogation Approach is based on leveraging our knowledge of the detain-ee's culture and mindset, together with using information we already know about him.... This Informed Interrogation Approach is in sharp contrast with the harsh interrogation approach introduced by outside contractors and forced upon CIA officials to use. ... The [enhanced interrogation] approach applies a force continuum, each time using harsher and harsher techniques until the detainee submits.... There are many problems with this tech-nique. A major problem is that it is ineffective. Al-Qaeda terrorists are trained to resist torture. As shocking as these techniques are to us, the al-Qaeda training prepares them for much worse—the torture they would expect to receive if caught by dictatorships for example. ... A second major problem with this technique is that evidence gained from it is unreliable. There is no way to know whether the detainee is being truthful, or just speak-ing to either mitigate his discomfort or to deliberately provide false information. ...
>
> Another disastrous consequence of the use of the harsh techniques was that it reintro-duced the 'Chinese Wall' between the CIA and FBI—similar to the wall that prevented us from working together to stop 9/11.... The situation was, and remains, too risky to allow someone to experiment with amateurish, Hollywood style interrogation methods—that in reality—taints sources, risks outcomes, ignores the end game, and diminishes our moral high ground in a battle that is impossible to win without first capturing the hearts and minds around the world. It was one of the worst and most harmful decisions made in our efforts against al-Qaeda.

U.S. senators Dianne Feinstein (Senate Intelligence Committee Chairman) and Carl Levin (Senate Armed Services Committee Chairman) released a joint statement in April 2012 saying that "We are deeply troubled by the claims of the CIA's former Deputy Director of Operations Jose Rodriguez regarding the effec-tiveness of the CIA's coercive interrogation techniques. We are also troubled by Rodriguez's statements justifying the destruction of video tapes documenting the use of coercive interrogation techniques as 'just getting rid of some ugly visuals.' His decision to order the destruction of the tapes was in violation of instructions from CIA to White House lawyers, illustrates a blatant disregard for the law, and unnecessarily caused damage to the CIA's reputation" [42].

The 2012 movie *Zero Dark Thirty* dramatized the hunt for bin Laden by the U.S. intelligence community. Acting CIA director Michael Morell issued a state-ment to agency employees on December 21, 2012 saying that *Zero Dark Thirty* is not a historically accurate film [43]. U.S. senators Dianne Feinstein, Carl Levin, and John McCain wrote in a letter to Sony Pictures Entertainment that "*Zero Dark Thirty* is factually inaccurate, and we believe that you have an obligation to state that the role of torture in the hunt for Usama Bin Laden is not based on the facts, but rather part of the film's fictional narrative. ... The filmmakers and your pro-duction studio are perpetuating the myth that torture is effective. You have a social and moral obligation to get the facts right" [44].

CNN national security analyst Peter Bergen also remarked that "the compelling story told in the film captures a lot that is true about the search for al-Qaeda's leader but also distorts the story in ways that could give its likely audience of millions of Americans the misleading picture that coercive interrogation techniques used by the CIA on al-Qaeda detainees—such as waterboarding, physical abuse and sleep deprivation—were essential to finding bin Laden" [45].

2.8 U.S. Fusion Centers

To encourage information sharing among governmental agencies, seventy-seven fusion centers have been set up in various cities across the U.S. "Fusion centers have stepped up to meet an urgent need in the last decade," Homeland Security and Governmental Affairs Committee Chairman Joe Lieberman said in October 2012. "Without fusion centers, we would not be able to connect the dots. Fusion centers have been essential to breaking down the information silos and communications barriers that kept the government from detecting the most horrific terrorist attack on this country—even though federal, state, and local officials each held valuable pieces of the puzzle" [46]. Lieberman cited examples of counterterrorism cases:

- "Raleigh Jihad" case—This case from 2009 involved Daniel Patrick Boyd and six others who planned to attack Marine Base Quantico. The North Carolina fusion center partnered with the local FBI Joint Terrorism Task Force on this investigation.
- Rezwan Ferdaus—In 2010, homegrown violent Islamist extremist Ferdaus was arrested in Boston for planning to attack the Pentagon and the Capitol with explosives attached to remote control small planes. The Massachusetts state fusion center was credited with making a "significant contribution" to the investigation.
- Seattle military recruiting center plot—In 2011, two homegrown violent Islamist extremists were arrested in Seattle for planning to attack a military recruiting center. The initial lead in this case came from a Seattle Police Department informant, and the investigation was jointly coordinated by the FBI and state and local agencies at the Washington state fusion center.

However, U.S. Senators Carl Levin and Tom Coburn issued a report in October 2012 of their two-year investigation that uncovered the ineffectiveness and inefficiency of the state and local fusion centers around the country [47]:

Sharing terrorism-related information between state, local and Federal officials is crucial to protecting the United States from another terrorist attack. Achieving this objective was the motivation for Congress and the White House to invest hundreds of millions of taxpayer dollars over the last nine years in support of dozens of state and local fusion centers across the United States.

The Subcommittee investigation found that DHS [Department of Homeland Security]-assigned detailees to the fusion centers forwarded "intelligence" of uneven quality—oftentimes shoddy, rarely timely, sometimes endangering citizens' civil liberties and

Privacy Act protections, occasionally taken from already-published public sources, and more often than not unrelated to terrorism.

The findings of both the 2010 and 2011 assessments contradict public statements by DHS officials who have described fusion centers as "one of the centerpieces of our counterterrorism strategy," and "a major force multiplier in the counterterrorism enterprise." The Subcommittee investigation found that the fusion centers often produced irrelevant, useless or inappropriate intelligence reporting to DHS, and many produced no intelligence reporting whatsoever.

Despite some success stories cited by Joe Lieberman, the Senate investigative report clearly reveals that the U.S. fusion centers are in dire need of revamp. Reorganization is a common practice in successful businesses, the U.S. government should learn from the private sector.

2.9 Hard Lessons from Pearl Harbor and 9/11

In an unclassified document released on September 2007, Tom Johnson, head of the Division of History and Publications at the NSA, analyzed the surprise Japanese attack on Pearl Harbor in 1941. Johnson wrote in *Cryptologic Quarterly* [48]:

Interservice and intraservice rivalry always loses. If the coordination between Army and Navy was bad in Washington and in Hawaii, it was even worse within the Navy itself. Admiral Turner unquestionably harmed the defense effort through overzealous aggrandizement and turf quarrels. It was inexcusable then—it is inexcusable today. And it can be seen everywhere one goes in the Federal Government.

There is a delicate balance between the requirements of secrecy and the needs of the customer. At Pearl Harbor this balance was not properly struck. Information was kept from field commanders on whose shoulders the administration had placed a great deal of responsibility. Information did not flow because we feared losing the source. It remained bottled up in Washington, serving as small talk for intelligence professionals, State Department officials, and a limited number of operational staff planners. It is not easy to achieve a balance, but it must be done, constantly, in thousands of daily decisions over disclosure and dissemination. We face the same decisions today, in far greater quantity, though with no greater consequence. We weren't smart about it then. Are we now?

After the surprise 9/11 attacks in similar magnitude to the attack on Pearl Harbor, the U.S. intelligence community has been moving in the right direction from "need-to-know" to "need-to-share" and "need-to-provide," but there is still plenty of room for improvement.

On September 11, 2012, terrorists attacked the U.S. Consulate in Benghazi, Libya during the massive protests across the Muslim world over an anti-Islam film [49]. U.S. ambassador J. Christopher Stevens and three other Americans were killed.

Fran Townsend, CNN national security analyst and former homeland security advisor to President George W. Bush, reported on the progress of the FBI investigation 15 days after the attack. Townsend said on CNN's "Anderson Cooper 360°," "They've gotten as far as Tripoli now, but they've never gotten to Benghazi," said Townsend. "They had difficulty, and we understand there was some bureaucratic infighting between the FBI and Justice Department on the one hand, and the State Department on the other, and so it took them longer than they would have liked to get into the country. They've now gotten there. But they still are unable to get permission to go to Benghazi" [50]. The FBI finally arrived at the crime scene on October 3, three weeks after the deadly attack on the U.S. Consulate [51].

On October 10, former regional security officer Eric Nordstrom testified before a congressional hearing that his superiors denied his request for additional security for the U.S. Consulate in Benghazi months before the terrorist attack. Nordstorm told the House Oversight Committee, "You know what makes it most frustrating about this assignment? It's not the hardships. It's not the gunfire. It's not the threats. It's dealing and fighting against the people, programs, and personnel who are supposed to be supporting me. ... For me, the Taliban is on the inside of the building" [52].

"Taliban on the inside of the State Department" is a dire metaphor. Bureaucratic red tape continues to hamper the effectiveness and efficiency of the U.S. intelligence community. The U.S. government needs to learn from successful private businesses that run an effective and efficient operation in serving their customers and outwitting their competitors. Like running any successful business, the government cannot afford to make serious mistakes. Bruce Riedel, former CIA officer and chair of Strategic Policy Review of Afghanistan and Pakistan, voiced his opinion in *The Daily Beast*, "In fighting terror, our team has to stay lucky 100 percent of the time. Al-Qaeda needs to be lucky only once" [53].

References

1. **Public Broadcasting Service.** Bin Laden's Fatwa. [Online] PBS Newshour, August 23, 1996. http://www.pbs.org/newshour/updates/military/july-dec96/fatwa_1996.html.
2. **PBS.** Al Qaeda's Fatwa. [Online] PBS Newshour, February 23, 1998. http://www.pbs.org/newshour/terrorism/international/fatwa_1998.html.
3. **Farnsworth, Elizabeth.** African Embassy Bombings. [Online] PBS Newshour, August 7, 1998. http://www.pbs.org/newshour/bb/africa/embassy_bombing/news_8-7-98.html.
4. **DCI Counterterrorist Center.** Bin Ladin Preparing to Hijack US Aircraft and Other Attacks. [Online] Central Intelligence Agency, December 4, 1998. http://www.foia.cia.gov/docs/DOC_0001110635/0001110635_0001.gif.
5. **The FBI.** FBI Ten Most Wanted Fugitive. [Online] Federal Bureau of Investigation, June 1999. http://www.fbi.gov/wanted/topten/usama-bin-laden.
6. **CNN.** Bin Laden says he wasn't behind attacks. [Online] CNN, September 16, 2001. http://articles.cnn.com/2001-09-16/us/inv.binladen.denial_1_bin-laden-taliban-supreme-leader-mullah-mohammed-omar.
7. Bin Laden on tape: Attacks 'all that we had hoped for'. [Online] CNN, December 13, 2001. http://articles.cnn.com/2001-12-13/us/ret.bin.laden.videotape_1_government-translation-bin-laden-talks-osama-bin.
8. **Dudney, Robert S.** Verbatim Special: War on Terror. [Online] AIR FORCE Magazine, December 2001. http://www.airforce-magazine.com/MagazineArchive/Documents/2001/December%202001/1201verb.pdf.
9. **ABC News.** Transcript of the Alleged Bin Laden Tape. [Online] ABC News, May 23, 2006. http://abcnews.go.com/International/Terrorism/story?id=1995630.

10. **Koch, Kathleen.** White House downplays Newsweek report. [Online] CNN, June 3, 2002. http://articles.cnn.com/2002-06-03/politics/white.house.newsweek_1_bin-laden-operatives-qaeda-bush-officials.

11. **Ensor, David.** Sources: CIA warned FBI about hijacker. [Online] CNN, June 4, 2002. http://articles.cnn.com/2002-06-03/us/cia.fbi.hijackers_1_almihdhar-and-nawaf-alhazmi-cia-analysts-al-qaeda-meeting.

12. **The Guardian.** Suspected Hijackers. [Online] The Guardian. http://www.guardian.co.uk/september11/suspectedhijackers/0,,605011,00.html.

13. **Wright, Lawrence.** The Agent. Did the C.I.A. stop an F.B.I. detective from preventing 9/11? [Online] The New Yorker, July 10, 2006. http://www.newyorker.com/archive/2006/07/10/060710fa_fact_wright?currentPage=all.

14. **The FBI.** The USS Cole Bombing. [Online] Federal Bureau of Investigation. http://www.fbi.gov/about-us/history/famous-cases/uss-cole.

15. **CNN.** New York reduces 9/11 death toll by 40. [Online] CNN, October 29, 2003. http://articles.cnn.com/2003-10-29/us/wtc.deaths_1_death-toll-world-trade-center-names.

16. **Roddy, Dennis B., et al.** Flight 93: Forty lives, one destiny. [Online] Pittsburgh Post-Gazette, October 28, 2001. http://old.post-gazette.com/headlines/20011028flt93mainstoryp7.asp.

17. **Benson, Pam.** Yemen plot exposes new world of U.S. spying. [Online] CNN, May 11, 2012. http://security.blogs.cnn.com/2012/05/11/yemen-plot-exposes-new-world of us-spying/.

18. **Scheuer, Michael F.** Bill and Dick, Osama and Sandy. [Online] The Washington Times, July 4, 2006. http://www.washingtontimes.com/news/2006/jul/4/20060704-110004-4280r/.

19. **Hersh, Seymour M.** What Went Wrong: The C.I.A. and the failure of American intelligence. [Online] The New Yorker, October 8, 2001. http://www.newyorker.com/archive/2001/10/08/011008fa_FACT?currentPage=all.

20. **Office of the Inspector General.** A Review of the FBI's Handling of Intelligence Information Prior to the September 11 Attacks. [Online] U.S. Justice Department, November 2004. http://www.justice.gov/oig/special/0506/chapter2.htm.

21. **Moynihan, Daniel Patrick.** Secrecy: The American Experience. [Online] Yale University Press, 1998. http://www.nytimes.com/books/first/m/moynihan-secrecy.html.

22. **The 9/11 Commission.** The System was Blinking Red. [Online] National Commission on Terrorist Attacks Upon the United States, July 22, 2004. http://www.9-11commission.gov/report/911Report_Ch8.htm.

23. **The 9-11 Commission.** The 9/11 Commission Report: Final Report of the National Commission on Terrorist Attacks Upon the United States (9/11 Report). [Online] U.S. Congress, July 22, 2004. http://www.9-11commission.gov/report/index.htm.

24. **Bamford, James.** War of Secrets; Eyes in the Sky, Ears to the Wall, and Still Wanting. [Online] The New York Times, September 8, 2002. http://www.nytimes.com/2002/09/08/weekinreview/war-of-secrets-eyes-in-the-sky-ears-to-the-wall-and-still-wanting.html?pagewanted=all.

25. **Block, Robert, Fields, Gary and Wrighton, Jo.** U.S. 'Terror' List Still Lacking. [Online] The Wall Street Journal, January 2, 2004. http://online.wsj.com/article/0,,SB107300534268021800,00.html.

26. **Investigation, Federal Bureau of.** Terrorist Screening Center. [Online] FBI. [Cited: November 23, 2012.] http://www.fbi.gov/about-us/nsb/tsc.

27. **Gorman, Siobhan.** How Team of Geeks Cracked Spy Trade. [Online] The Wall Street Journal, September 4, 2009. http://online.wsj.com/article/SB125200842406984303.html.

28. **Rudy, Charles J.** One team, one fight. [Online] U.S. Air Force (Kandahar Airfield), March 16, 2011. http://www.kdab.afcent.af.mil/news/story.asp?id=123246976.

29. **GeekyRoom's Chief Editor.** CIA's Open Source Center Analyze Twitter in Real Time. [Online] GeekyRoom, November 5, 2011. http://geekyroom.com/2011/11/05/cia-open-source-center-analyze-twitter-in-real-time/.

30. **Open Source Center.** Open Source Center: Information to Intelligence. [Online] U.S. Government. https://www.opensource.gov/.

31. **108th Congress.** Intelligence Reform and Terrorism Prevention Act of 2004. [Online] National Counterterrorism Center, December 17, 2004. http://www.nctc.gov/docs/pl108_458.pdf.
32. **Rose, Derek.** Cia Hid Key Info On 9/11. [Online] New York Daily News, July 3, 2006. http://articles.nydailynews.com/2006-07-03/news/18345695_1_two-al-qaeda-cia-cole-bombing.
33. **Office of the Director of National Intelligence.** 100 Day Plan for Integration and Collaboration. [Online] Defense Technical Information Center (DTIC), September 2007. http://www.dtic.mil/dtic/tr/fulltext/u2/a471965.pdf.
34. **Director of National Intelligence.** United States Intelligence Community 500 Day Plan Integration and Collaboration. [Online] Defense Technical Information Center (DTIC), October 10, 2007. http://www.dtic.mil/dtic/tr/fulltext/u2/a473641.pdf.
35. **Office of the Director of National Intelligence (ODNI).** Office of the Director of National Intelligence: Leading Intelligence Integration. [Online] U.S. Government. http://www.dni.gov/.
36. **Wright, Lawrence.** The Spymaster. Can Mike McConnell fix America's intelligence community? [Online] The New Yorker, January 21, 2008. http://www.newyorker.com/reporting/2008/01/21/080121fa_fact_wright?currentPage=all .
37. **Young, Denise.** Letitia Long: A Global Vision. Alumna leads intelligence agency in new era of collaboration. [Online] Virginia Tech Magazine, Spring 2012. http://www.vtmag.vt.edu/spring12/letitia-long.html.
38. **Hawkins, Arielle.** Bin Laden raid nets one intel employee big bonus. [Online] CNN, May 18, 2012. http://security.blogs.cnn.com/2012/05/18/bin-laden-raid-nets-one-intel-employee-big-bonus/.
39. **Crawford, Jamie.** McRaven on bin Laden raid: One of history's "great intelligence operations". [Online] CNN, July 26, 2012. http://security.blogs.cnn.com/2012/07/26/u-s-special-ops-commander-discusses-role-in-serious-and-humorous-tones/.
40. **Feinsten, Dianne and Levin, Carl.** Feinstein, Levin Statement on CIA's Coercive Interrogation Techniques. [Online] Dianne Feinstein: U.S. Senator for California, April 30, 2012. http://www.feinstein.senate.gov/public/index.cfm/2012/4/feinstein-levin-statement-on-cia-s-coercive-interrogation-techniques.
41. **Soufan, Ali.** Testimony of Ali Soufan. [Online] United States Senate Committee on the Judiciary, May 13, 2009. http://www.judiciary.senate.gov/hearings/testimony.cfm?id=e655f9e2809e5476862f735da14945e6&wit_id=e655f9e2809e5476862f735da14945e6-1-2.
42. **Feinstein, Dianne and Levin, Carl.** Feinstein, Levin Statement on CIA's Coercive Interrogation Techniques. [Online] United States Senate, April 30, 2012. http://www.feinstein.senate.gov/public/index.cfm/press-releases?ID=f3271910-3fad-40a5-9d98-93450e0090aa.
43. **Benson, Pam.** CIA challenges accuracy of 'Zero Dark Thirty'. [Online] CNN, December 21, 2012. http://security.blogs.cnn.com/2012/12/21/cia-challenges-accuracy-of-zero-dark-thirty/.
44. **Feinstein, Dianne; Levin, Carl; McCain John.** Feinstein Releases Statement on 'Zero Dark Thirty'. [Online] United States Senate, December 19, 2012. http://www.feinstein.senate.gov/public/index.cfm/press-releases?ID=b5946751-2054-404a-89b7-b81e1271efc9.
45. **Bergen, Peter.** 'Zero Dark Thirty': Did torture really net bin Laden? [Online] CNN, December 10, 2012. http://www.cnn.com/2012/12/10/opinion/bergen-zero-dark-thirty/index.html.
46. **Lieberman, Joe.** Fusion Centers Add Value to Federal Government Counterterrorism Efforts. [Online] U.S. Senate Committee on Homeland Security and Governmental Affairs, October 3, 2012. http://www.hsgac.senate.gov/media/fusion-centers-add-value-to-federal-government-counterterrorism-efforts.
47. **Levin, Carl and Coburn, Tom.** Federal Support for and Involvement in State and Local Fusion Centers. [Online] U.S. Senate Committee on Homeland Security and Governmental Affairs, October 3, 2012. http://www.hsgac.senate.gov/download/?id=49139e81-1dd7-4788-a3bb-d6e7d97dde04.
48. **Johnson, Tom.** What Every Crytologist Should Know about Pearl Harbor. [Online] Cryptologic Quarterly, September 27, 2007. http://www.nsa.gov/public_info/_files/cryptologic_quarterly/pearlharbor.pdf.

49. **Reuters.** Terrorists killed U.S. ambassador to Libya: Panetta. [Online] Chicago Tribune, September 27, 2012. http://www.chicagotribune.com/news/sns-rt-us-libya-usa-investigation-bre88q1jw-20120927,0,3904573.story.

50. **CNN Wire Staff.** Sources: 15 days after Benghazi attack, FBI still investigating from afar. [Online] CNN, September 26, 2012. http://www.cnn.com/2012/09/26/world/africa/libya-investigation/index.html.

51. **Starr, Barbara.** FBI visits site of attack in Libya. [Online] CNN, October 4, 2012. http://www.cnn.com/2012/10/04/world/africa/libya-fbi-benghazi/index.html.

52. **CNN Wire Staff.** U.S. official says superiors worked against effort to boost Benghazi security. [Online] CNN, October 11, 2012. http://www.cnn.com/2012/10/10/politics/congress-libya-attack/index.html.

53. **Riedel, Bruce.** A Stubborn Terror. [Online] The Daily Beast, September 10, 2012. http://www.thedailybeast.com/newsweek/2012/09/09/a-stubborn-terror.html.

Part II
Counterterrorism Technologies: Total Information Awareness and Data Mining

Chapter 3
The Rise and Fall of Total Information Awareness

Information is the oxygen of the modern age.
—President Ronald Reagan (June 14, 1989).

It would be no good to solve the security problem and give up the privacy and civil liberties that make our country great.
—Admiral John Poindexter (August 12, 2003).

Scientia potentia est. (Knowledge is power).
—Thomas Hobbes in *De Homine (Man)* (1658).

3.1 President Ronald Reagan and Admiral John Poindexter

President Ronald Reagan had long recognized the vital importance of communications technology and information sharing as he said in June 1989 after having served two terms as the President of the United States [1]:

Information is the oxygen of the modern age. ... It seeps through the walls topped with barbed wire. It wafts across the electrified, booby-trapped borders. Breezes of electronic beams blow through the Iron Curtain as if it was lace. ... The Goliath of totalitarian control will rapidly be brought down by the David of the microchip.

Back in April 1984, President Reagan signed the National Security Decision Directive (NSDD) 138: Combating Terrorism, which authorized the increase of intelligence collection directed against groups or states involved in terrorism [2].

Reagan appointed Navy Vice Admiral John Poindexter as the National Security Advisor in December 1985. With a Ph.D. in Nuclear Physics from California Institute of Technology (Caltech), Poindexter had been a strong advocate of new computer technology and distributed data management system during his tenure in the U.S. military. In November 1986, however, Poindexter was forced to resign from the White House Office and retire as Rear Admiral due to his role in the Iran-Contra Affair [3].

N. Lee, *Counterterrorism and Cybersecurity*, DOI: 10.1007/978-1-4614-7205-6_3,
© Springer Science+Business Media New York 2013

After a 3-month investigation by the Tower Commission headed by former Senator John Tower, President Reagan addressed the nation in March 1987 acknowledging the danger of unchecked covert operations and the need for stronger presidential oversight [4]:

> A few months ago I told the American people I did not trade arms for hostages. My heart and my best intentions still tell me that's true, but the facts and the evidence tell me it is not. ... I'm taking action in three basic areas: personnel, national security policy, and the process for making sure that the system works. ... I have had issued a directive prohibiting the National Security Council (NSC) staff itself from undertaking covert operations—no ifs, ands, or buts.

In March 1988, Poindexter and Lieutenant Colonel Oliver North were indicted on charges of conspiracy to defraud the United States by illegally providing the Nicaraguan rebels with profits from the sale of American weapons to Iran [5]. In April 1990, Poindexter was convicted on five counts of lying to Congress and obstructing the Congressional investigation of the Reagan Administration's covert arms sales to Iran and the diversion of some proceeds to rebels fighting the Marxist Government in Nicaragua. However, in November 1991, the District of Columbia Circuit Court overturned Poindexter's conviction by a vote of two to one [6].

A day after September 11, 2001, Poindexter lamented with his close friend Brian Sharkey that they had not prevented the terrorist attacks [7]. Sharkey was a former program manager at the Defense Advanced Research Projects Agency (DARPA). Poindexter was working for BMT Syntek Technologies, a defense contractor that was developing Project Genoa, a data-mining decision-support system for DARPA. Genoa provided analyst tools to augment human cognitive processes and aid understanding of complex arguments [8]. After the 9/11 attacks, Poindexter wanted to put Project Genoa on steroids.

3.2 Defense Advanced Research Projects Agency

Defense Advanced Research Projects Agency (DARPA) was created as Advanced Research Projects Agency (ARPA) in 1958 by President Dwight Eisenhower in response to the surprise Sputnik launch by the Soviet Union a year before [9].

Since its formation, DARPA has made significant contributions to science and technology in collaboration with universities and research organizations. J. C. R. Licklider, director of DARPA's Information Processing Techniques Office (IPTO) in 1962, was one of the fathers of modern computer science such as the graphical user interface and personal workstations [10]. Licklider's vision of an interactive worldwide computer network led to the creation of Advanced Research Projects Agency Network (ARPANET), the predecessor to the Internet [11]. DARPA has been involved in many research and development projects such as the first satellite positioning system, stealth technology, next-generation supercomputers, and alternative energy [12].

Former DARPA director Regina Dugan spoke at TED 2012 in Long Beach, California where she introduced a robotic hummingbird, a prosthetic arm controlled by thought, and other DARPA inventions. Dugan said, "Scientists and engineers changed the world. I'd like to tell you about a magical place called DAPRA where scientists and engineers defy the impossible and refuse to fear failure" [13].

3.3 Information Awareness Office (IAO)

In January 2002, retired Admiral John Poindexter returned to the U.S. government to serve as the director of the newly established Information Awareness Office (IAO) at the Defense Advanced Research Projects Agency (DARPA) of the U.S. Department of Defense (DoD) [14].

Dr. Tony Tether, director of DARPA, established the IAO whose official seal featured a pyramid topped with an all-seeing eye, similar to the one on the back of a U.S. dollar bill. The seal also had a Latin inscription, SCIENTIA EST POTENTIA, which means that science has a lot of potential, or in other words, knowledge is power. (See Fig. 3.1).

IAO's mission was to "develop new tools to detect, anticipate, train for, and provide warnings about potential terrorist attacks" [14]. In his April 2002 statement to the U.S. Senate Committee on Armed Services, Tether explained the rationale behind the creation of the IAO to "find, identify, track, and understand terrorist networks" [15]:

> One of the great challenges in the war on terrorism is to know our enemy—who he is, where he is, and what he's doing. In order to focus our efforts, I established another new DARPA office, the Information Awareness Office (IAO). IAO is developing the information systems needed to find, identify, track, and understand terrorist networks and vastly improve what we know about our adversaries. We will use the light of information technology to take away the shadows they hide in.
>
> For example, IAO's Evidence Extraction and Link Discovery program is aimed at finding terrorist networks hidden in the mountains of diverse data that we collect. The Wargaming the Asymmetric Environment program is explicitly aimed at predicting the behavior of terrorist groups in some detail, an extremely difficult challenge. Usually what we do now is issue broad warnings to the public to be on guard, like the several that were announced following September 11th. Wargaming the Asymmetric

Fig. 3.1 Official seal of the information awareness office (*IAO*)

Environment seeks to move from those broad warnings to more specific predictions. In short, we want to go from predicting the terrorist "climate" to predicting the terrorist "weather." Some would argue that this is an outrageous goal, one that is not possible to achieve. I agree it sounds outrageous, but what if we can do it? That is why it is a DARPA program.

In addition, IAO's **Total Information Awareness** program is now setting up a testbed at the Army's Intelligence and Security Command to test our new technologies on real-world threat data.

3.4 Perception of Privacy Invasion

Although Dr. Tony Tether's intention for Total Information Awareness (TIA) was to prevent future terrorist attacks, the American public was startled by *The New York Times* headline on November 9, 2002: "Pentagon Plans a Computer System That Would Peek at Personal Data of Americans." In his article, senior writer John Markoff expressed his deep concerns about privacy invasion by the U.S. government [16]:

The Pentagon is constructing a computer system that could create a vast electronic dragnet, searching for personal information as part of the hunt for terrorists around the globe—including the United States.

As the director of the effort, Vice Adm. John M. Poindexter, has described the system in Pentagon documents and in speeches, it will provide intelligence analysts and law enforcement officials with instant access to information from Internet mail and calling records to credit card and banking transactions and travel documents, without a search warrant.

Historically, military and intelligence agencies have not been permitted to spy on Americans without extraordinary legal authorization. But Admiral Poindexter, the former national security adviser in the Reagan administration, has argued that the government needs broad new powers to process, store and mine billions of minute details of electronic life in the United States.

In order to deploy such a system, known as **Total Information Awareness**, new legislation would be needed. ... That legislation would amend the Privacy Act of 1974, which was intended to limit what government agencies could do with private information.

Senior editor Hendrik Hertzberg of *The New Yorker* concurred with Markoff in his December 9, 2002 article comparing the Information Awareness Office with Dr. Strangelove's vision [17]:

The [Information Awareness] Office's main assignment is, basically, to turn everything in cyberspace about everybody—tax records, driver's license applications, travel records, bank records, raw FBI files, telephone records, credit card records, shopping mall security camera videotapes, medical records, every e-mail anybody ever sent—into a single, humongous, multi-googolplexibyte database that electronic robots will mine for patterns of information suggestive of terrorist activity. Dr. Strangelove's vision—"a chikentic gomplex of gumbyuders"—is at last coming into its own.

"This could be the perfect storm for civil liberties in America," said Marc Rotenberg, president and executive director of the Electronic Privacy Information Center (EPIC) in Washington. "The vehicle is the Homeland Security Act, the technology is DARPA and the agency is the FBI. The outcome is a system of national surveillance of the American public" [16].

3.5 Privacy Protection in Total Information Awareness (TIA)

Since the establishment of the Information Awareness Office in January 2002, Dr. Tony Tether was aware of potential privacy issues. In March 2002, DARPA issued a formal call for proposals (BAA02–08) on Information Awareness [18]. The purpose was to "develop information technologies to help prevent continued terrorist attacks on the citizens, institutions, and property of the United States and its allies." Information repositories and privacy protection technologies were the number 1 priority on the three stated objectives:

1. Development of revolutionary technology for ultra-large all-source information repositories and associated **privacy protection technologies**;
2. Development of collaboration, automation, and cognitive aids technologies that allow humans and machines to think together about complicated and complex problems more efficiently and effectively; and
3. Development and implementation of an end-to-end, closed-loop prototype system to aid in countering terrorism through prevention by integrating technology and components from existing DARPA programs such as: Genoa, Evidence Extraction and Link Discovery (EELD), Wargaming the Asymmetric Environment (WAE), Translingual Information Detection, Extraction and Summarization (TIDES), Human Identification at Distance (HID), Bio-Surveillance; as well as programs resulting from the first two areas of this BAA and other programs.

Figure 3.2 is a diagram of the Total Information Awareness (TIA) system designed by the Information Awareness Office (IAO) showing the workflow from detection, classification, identification, tracking, understanding, to preemption. "Privacy and Security" is highlighted in red, filtering the data needed for repositories and counterterrorism analysis.

To accommodate "ultra-large all-source information," the database envisioned is "of an unprecedented scale, will most likely be distributed, must be capable of being continuously updated, and must support both autonomous and semi-automated analysis. The latter requirement implies that the representation used must, to the greatest extent possible, be interpretable by both algorithms and human analysts. The database must support change detection and be able to execute automated procedures implied by new information."

DARPA acknowledged that "the reduced signature and misinformation introduced by terrorists who are attempting to hide and deceive imply that uncertainty must be represented in some way." The call for proposals stressed the importance "to protect the privacy of individuals not affiliated with terrorism" by seeking "technologies for controlling automated search and exploitation algorithms and for purging data structures appropriately."

Fig. 3.2 Total information awareness of transnational threats requires keeping track of individuals and understanding how they fit into models (Courtesy of the Defense Advanced Research Projects Agency)

In fact, the TIA program included the Genisys Privacy Protection Program. The goal of Genisys was to make databases easy to use and simple to integrate. The Genisys privacy protection program would ensure personal privacy and protect sensitive intelligence sources [19]:

> Information systems and databases have the potential to identify terrorist signatures through the transactions they make, but Americans are rightfully concerned that data collection, integration, analysis, and mining activities implicate privacy interests. The Genisys Privacy Protection Program aims to provide security with privacy by providing certain critical data to analysts while controlling access to unauthorized information, enforcing laws and policies through software mechanisms, and ensuring that any misuse of data can be quickly detected and addressed. Research being conducted under other IAO programs may indicate that information about terrorist planning and preparation activities exists in databases that also contain information about U.S. persons. Privacy protection technologies like those being developed under the Genisys Privacy Protection Program would be essential to protect the privacy of U.S. citizens should access to this sort of information ever be contemplated.

Barbara Simons, computer scientist and past president of the Association for Computing Machinery (ACM), was nonetheless highly skeptical: "I'm just not convinced that the TIA will give us tools for catching terrorists that we don't already have or that could be developed with far less expensive and less intrusive systems" [20].

On the particular issue of database security and privacy, Simons said, "Even if one were able to construct a system which did protect privacy in some sense, we certainly have not been very successful with building humongous databases that are secure. ... A lot of my colleagues are uncomfortable about this and worry about the potential uses that this technology might be put to, if not by this administration then by a future one. Once you've got it in place you can't control it" [16].

The economist warned of an Orwellian future in its 2003 special report on the Internet society: "As more human interactions are conducted and recorded electronically, as the ability to analyse databases grows and as video and other offline surveillance technologies become cheaper and more effective, it will become ever easier for authoritarian governments to set up systems of widespread surveillance. George Orwell's Big Brother of '1984' might yet become a reality, a few decades later than he expected" [21].

3.6   Opposing Views on TIA

In response to continuing criticism and public uproar, Dr. Tony Tether testified before the U.S. House of Representatives House Armed Services Committee in March 2003 [22]:

> The goal of our Information Awareness program is to create information technology that America's national security community can use to detect and defeat terrorist networks before they can attack us.
>
> One of our Information Awareness programs is Total Information Awareness (TIA), around which there has been much controversy. If I knew only what I read in the press about TIA, I would be concerned too. So I'd like to briefly address some of the main concerns.
>
> No American's privacy has changed in any way as a result of DARPA's Information Awareness programs, including the TIA. The Department of Defense *is not* developing technology so it can maintain dossiers on every American citizen. The Department of Defense *is not* assembling a giant database on Americans.
>
> Instead, the TIA program is designed as an experimental, multi-agency prototype network that participating agencies can use to better share, analyze, understand, and make decisions based on whatever data to which they currently have *legal* access. ITA will integrate three broad categories of information technologies from DARPA and elsewhere: advanced collaboration and decision support tools, language translation, and data search and pattern recognition.
>
> On February 7, 2003, the DoD announced the establishment of two boards to oversee TIA. These boards, an internal oversight board and an outside advisory committee, will work with DARPA as we continue our research. They will help ensure that TIA develops and disseminates its tools to track terrorists in a manner consistent with U.S. constitutional law, U.S. statutory law, and American values.

Tether and Poindexter believed that the 9/11 attacks could have been averted if intelligence information was collected and analyzed in time, especially about the known al-Qaeda terrorists Khalid al-Mihdhar and Nawaf al-Hazmi who entered the United States in January 2000.

Former *Newsweek* investigative correspondent Michael Isikoff said, "What's stunning is that, from that moment on, they [al-Mihdhar and al-Hazmi] lived entirely out in the open. They opened up bank accounts, they got a California driver's license, they opened up credit cards and they interacted with at least five other of the hijackers on 9/11" [23].

Proponents of total information awareness believe that TIA can eliminate the element of surprise so that we can prevent future 9/11 and Pearl Harbor attacks. Opponents disagree. Amy Belasco, specialist in national defense at the Foreign Affairs, Defense, and Trade Division, summarized the opposing views of TIA in her March 2003 report [14]:

> To proponents, TIA R&D holds out the promise of developing a sophisticated system that would develop new technologies to find patterns from multiple sources of information in order to give decision makers new tools to use to detect, preempt and react to potential terrorist attacks.
>
> To opponents, TIA has the potential to violate the privacy of individuals by giving the government access to vast amounts of information about individuals as well as possibly misidentifying individuals as potential terrorists.

3.7 Demystifying IAO and TIA

The Total Information Awareness (TIA) program involves multiple research and development (R&D) programs and the integration of these programs into a prototype TIA system. In July 2002, the Information Awareness Office (IAO) published a TIA System Description Document [24], and DARPA dedicated $137.5 million to the R&D programs and $10 million to the system integration for fiscal year 2003 [14].

In April 2003, the IAO issued an elaborate, year-long call for proposals (BAA03–23) on Information Awareness [25]. The IAO was soliciting ideas that "will imagine, develop, apply, integrate, demonstrate and transition information technologies and components for possible use in prototype closed-loop information systems to counter asymmetric threats." The proposal explained the goal of the Total Information Awareness program [25]:

> Program outputs will exploit information to significantly improve preemption capabilities, national security warning, and national security decision-making. The most serious asymmetric threat facing the United States is terrorism, a threat characterized by collections of people loosely organized in shadowy networks that are difficult to identify and define. IAO plans to develop technology that will allow understanding of the intent of these networks, their plans, and potentially define opportunities for disrupting or eliminating the threats. To effectively and efficiently carry this out, we must promote sharing, collaborating and reasoning to convert nebulous data to knowledge and actionable options. IAO will accomplish this by pursuing the development of technologies, components, and applications that may become integrated in a prototype [Total Information Awareness] system.

The IAO had no interest in developing information collection technology. Instead, its primary interest was in the following topic areas [26]:

1. Collaborative Reasoning and Decision Support Technologies.
   - Detect terrorist planning and preparation activities.
   - Facilitate information sharing.
   - Conduct simulation and risk analysis.

- Investigate structured argumentation, evidential reasoning, storytelling, change detection, and truth maintenance.

2. Language Translation Technologies.
   - Detect, extract, summarize, and translate information.
   - Develop speech-to-text transcription technologies for English, Chinese, and Arabic languages.
   - Port applications to new languages within 1 month.

3. Pattern Recognition and Predictive Modeling Technologies.
   - Extract evidence and find patterns from vast amounts of unstructured textual data (such as intelligence messages or news reports that are legally available and obtainable by the U.S. Government).
   - Discover critical information from speech and text of multiple languages.
   - Develop threat-specific tools to enable analysts and decision makers to predict terrorist attacks and to simulate potential intervention strategies.
   - Identify abnormal health detectors indicative of a biological attack.

4 Data Search and Privacy Protection Technologies.
   - Exploit distributed databases, information repositories, and sensor feeds.
   - Represent uncertainty in structured data.
   - Develop privacy protection technologies including immutable audit, self-reporting data, tamper-proof accounting system, anonymization and inferencing techniques, use of filtering and expunging software agents, and selective revelation concepts.

5. Biometric Technologies.

   - Develop automated, multimodal, biometric technologies to detect, recognize, and identify humans, alone or grouped, in disguise or not, at a distance, day or night, and in all weather conditions.
   - Investigate 3D morphable modeling approaches, the feasibility of networking and fusing multiple biometric sensors, and activity recognition monitoring concepts.

Figure 3.3 shows the overall organization and activities of the Information Awareness Office (IAO) dated May 2003 [19]. The IAO was responsible for transitioning appropriate technologies to the Total Information Awareness (TIA) system from R&D programs in advanced collaboration and decision support, language translation, data search, pattern recognition, and privacy protection. The TIA would be hosted by the Information Operations Center at Intelligence and Security Command (U.S. Army's INSCOM).

Figure 3.4 is a TIA reference model using a signal processing analogy to show how the software components from IAO and other government programs and from commercial sources fit together. TIA provides the analysts with the capability to discover the plans and intentions of potential terrorist activities by building and refining models of terrorist attacks based on available information.

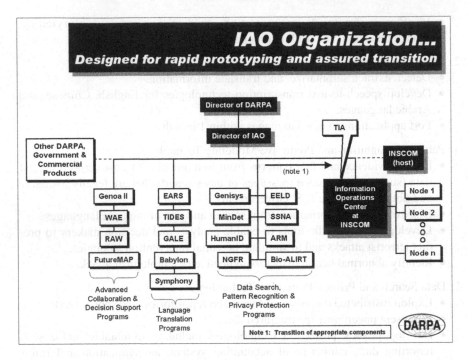

Fig. 3.3 Information Awareness Office (*IAO*) (Courtesy of the Defense Advanced Research Projects Agency)

3.8 Demise of IAO and TIA

Despite the support of the Bush administration and the last-ditched effort by IAO renaming Total Information Awareness to Terrorism Information Awareness in May 2003 [19], adverse media reaction and public distrust of Admiral John Poindexter proved to be too strong to overcome.

In July 2003, Poindexter faced harsh criticism from the media about IAO's Policy Analysis Market (PAM), part of the Futures Markets Applied to Prediction (FutureMAP) project. PAM was an online futures trading market in which anonymous speculators would bet on forecasting terrorist attacks, assassinations, and coups d'état [27]. IAO justified PAM by stating that such futures trading had proven effective in predicting other events like oil prices, elections, and movie ticket sales. Business journalist James Surowiecki of *The New Yorker* remarked, "That's especially important in the case of the intelligence community because we know that, for example, in the case of 9/11 there was lots of valuable and relevant information available before the attack took place. What was missing was a mechanism for aggregating that information in a single place. A well-designed market might have served as that mechanism" [28].

However, a sample bet on the assassination of Yasser Arafat proved to be simply unacceptable to the U.S. Congress [29]. Many U.S. Senators rebuked

Fig. 3.4 Total Information Awareness (*TIA*) reference model (Courtesy of the Defense Advanced Research Projects Agency)

Poindexter for applying economic theory of efficient markets and market discovery to national security [30]:

> Hillary Clinton of New York: "It's a futures market on death, and not in keeping with our values."
>
> Tom Daschle of South Dakota: "I couldn't believe that we would actually commit $8 million to create a Web site that would encourage investors to bet on futures involving terrorist attacks and public assassinations."
>
> Ron Wyden of Oregon: "The idea of a federal betting parlor on atrocities and terrorism is ridiculous and it's grotesque."
>
> Byron Dorgan of North Dakota: "Can you imagine if another country set up a betting parlor so that people could go in and bet on the assassination of an American political figure?"

FutureMAP and PAM were the last straw for Poindexter, IAO, and TIA. A month later in August 2003, Poindexter resigned with an open letter to DARPA director Tony Tether, in which he vehemently defended his actions and viewpoints [31]:

> [On first premise/research path—TIA:]
>
> As you know as our research has evolved we have had basically two research paths— each in the context of a premise. The first premise is that the U.S. government has all of the data it needs to find information that would allow us to detect foreign terrorists and their plans and thus enable the prevention of attacks against U.S. interests. ... On this first research path we created an experimental network called TIA and partnered with nine foreign intelligence, counter-intelligence and military commands for testing experimental

tools using foreign intelligence data that is currently available to them. ... The work under this premise should not be controversial in the U.S. since the tools are being applied using foreign intelligence data and as I have said is completely responsive to the problems the Congress has raised with respect to 9/11.

[On second premise/research path—FutureMAP:]

If we are wrong on the first premise and the U.S. government does not have all of the data it needs to find the terrorists and prevent their attacks, we felt it prudent to explore a second research path. This is the controversial one. In terms of the recent flap over FutureMap—did we want to bet the safety of thousands if not millions of Americans that our first premise was correct? Since we didn't want to make that bet, we devoted a relatively small portion of the funds that had been made available to us to this second research path.

[On privacy issue:]

We knew from the beginning that this second research path would be controversial and if the research proved successful, we would have to solve the privacy issue if it were ever to be deployed. We did not want to make a tradeoff between security and privacy. It would be no good to solve the security problem and give up the privacy and civil liberties that make our country great. ... We needed to find a solution for all three concerns: privacy of US citizens, privacy of foreign citizens and privacy of sources and methods.

In early 2002, shortly after the new office was formed, we began a study called Security with Privacy to imagine ways technology could be developed to preserve the privacy of individuals and still search through data that is not currently available to the government looking for specific patterns of activity that are related to terrorist planning and preparation activities.

[On portrayal by major media:]

In November 2002 after our work had been badly misrepresented in the major media, it was decided that I should not speak publicly to provide a defense and explanation of our work since I was such a "lightning rod" (not my words). ... I regret we have not been able to make our case clear and reassure the public that we do not intend to spy on them. ...

[On closing plea:]

In my opinion, the complex issues facing this nation today may not be solved using historical solutions and rhetoric that has been applied in the past, and that it may be useful to explore complex solutions that sometimes involve controversial technical concepts in order to rediscover the privacy foundations of this nation's strength and the basis for its freedoms.

Poindexter's plea fell on deaf ears. A month later in September 2003, the U.S. Congress axed the Information Awareness Office and terminated the funding for TIA: "The conferees agree with the Senate position which eliminates funding for the Terrorism Information Awareness (TIA) program within the Defense Advanced Research Projects Agency (DARPA). The conferees are concerned about the activities of the Information Awareness Office (IAO) and direct that the Office be terminated immediately. The only research projects previously under the jurisdiction of the Information Awareness Office (IAO) that may continue under DARPA are: Bio-Event Advanced Leading Indicator Recognition Technology, Rapid Analytic Wargaming, Wargaming the Asymmetric Environment, and Automated Speech and Text Exploitation in Multiple Languages (including Babylon and Symphony)" [32].

References

1. **Associated Press.** Reagan Urges 'Risk' on Gorbachev : Soviet Leader May Be Only Hope for Change, He Says. [Online] Los Angeles Times, June 13, 1989. http://articles.latimes.com/1989-06-13/news/mn-2300_1_soviets-arms-control-iron-curtain.
2. **Office of Science and Technology Policy.** Obama Administration Unveils "Big Data" Initiative: Announces $200 Million In New R&D Investments. [Online] Executive Office of the President, March 29, 2012. http://www.whitehouse.gov/sites/default/files/microsites/ostp/big_data_press_release.pdf.
3. **The New York Times.** IRAN-CONTRA REPORT; Arms, Hostages and Contras: How a Secret Foreign Policy Unraveled. [Online] The New York Times, November 19, 1987. http://www.nytimes.com/1987/11/19/world/iran-contra-report-arms-hostages-contras-secret-foreign-policy-unraveled.html.
4. **Reagan, Ronald.** Address to the Nation on Iran-Contra. [Online] University of Virginia Miller Center, March 4, 1987. http://millercenter.org/scripps/archive/speeches/detail/3414.
5. **Shenon, Philip.** North, Poindexter and 2 Others Indicted on Iran-Contra Fraud and Theft Charges. [Online] The New York Times, March 17, 1988. http://www.nytimes.com/1988/03/17/world/north-poindexter-and-2-others-indicted-on-iran-contra-fraud-and-theft-charges.html?pagewanted=all.
6. **Greenhouse, Linda.** Supreme Court Roundup; Iran-Contra Appeal Refused by Court. [Online] The New York Times, December 8, 1992. http://www.nytimes.com/1992/12/08/us/supreme-court-roundup-iran-contra-appeal-refused-by-court.html.
7. **Harris, Shane.** Lightning Rod. [Online] Government Executive, July 15, 2004. http://www.govexec.com/magazine/features/2004/07/lightning-rod/17199/.
8. **DARPA.** Genoa. [Online] mirror of decommissioned Federal government site www.darpa.mil/iao/Genoa.htm. http://infowar.net/tia/www.darpa.mil/iao/Genoa.htm.
9. **Defense Advanced Research Projects Agency.** History. [Online] Defense Advanced Research Projects Agency. [Cited: November 29, 2012.] http://www.darpa.mil/about/history/history.aspx.
10. **Waldrop, Mitchell M.** No, This Man Invented The Internet. [Online] Forbes, November 27, 2000. http://www.forbes.com/asap/2000/1127/105.html.
11. **Van Atta, Richard.** 50 Years of Bridging the Gap. [Online] Defense Advanced Research Projects Agency. [Cited: November 29, 2012.] http://www.darpa.mil/WorkArea/DownloadAsset.aspx?id=2553.
12. **Defense Advanced Research Projects Agency.** First 50 Years. [Online] Defense Advanced Research Projects Agency. [Cited: November 29, 2012.] http://www.darpa.mil/About/History/First_50_Years.aspx.
13. **Dugan, Regina.** Regina Dugan: From mach-20 glider to humming bird drone. [Online] TED, March 2012. http://www.ted.com/talks/regina_dugan_from_mach_20_glider_to_humming_bird_drone.html.
14. **Belasco, Amy.** Total Information Awareness Programs: Funding, Composition, and Oversight Issues. [Online] The Air University, March 21, 2003. http://www.au.af.mil/au/awc/awcgate/crs/rl31786.pdf.
15. **Tether, Tony.** Statement by Dr. Tony Tether to the U.S. Senate Committee on Armed Services. [Online] U.S. Senate Committee on Armed Services, April 10, 2002. http://www.armed-services.senate.gov/statemnt/2002/April/Tether.pdf.
16. **Markoff, John.** Pentagon Plans a Computer System That Would Peek at Personal Data of Americans. [Online] The New York Times, November 9, 2002. http://www.nytimes.com/2002/11/09/politics/09COMP.html?pagewanted=all.
17. **Hertzberg, Hendrik.** Too Much Information. [Online] The New Yorker, December 9, 2002. http://www.newyorker.com/archive/2002/12/09/021209ta_talk_hertzberg.
18. **Defense Advanced Research Projects Agency.** INFORMATION AWARENESS Solicitation Number: BAA02-08. [Online] Federal Business Opportunities, March 21, 2002.

https://www.fbo.gov/index?s=opportunity&mode=form&tab=core&id=82cd202bb8a0528c
389f11d00dad8514&_cview=0.

19. **Electronic Privacy Information Center.** Report to Congress regarding the Terrorism Information Awarenss. [Online] May 20, 2003. http://epic.org/privacy/profiling/tia/may03_report.pdf.

20. **Manjoo, Farhad.** Total Information Awareness: Down, but not out. [Online] Salon, January 29, 2003. http://www.salon.com/2003/01/29/tia_privacy/.

21. **The Economist.** Caught in the net. [Online] The Economist, January 23, 2003. http://www.economist.com/node/1534249.

22. **Tether, Tony.** Statement by Dr. Tony Tether to the U.S. House of Representatives House Armed Services Committee. [Online] DARPA, March 27, 2003. www.darpa.mil/WorkArea/DownloadAsset.aspx?id=1778.

23. **Koch, Kathleen.** White House downplays Newsweek report. [Online] CNN, June 3, 2002. http://articles.cnn.com/2002-06-03/politics/white.house.newsweek_1_bin-laden-operatives-qaeda-bush-officials.

24. **Gregory, Mack.** Total Information Awareness Program (TIA) System Description Document (SDD). [Online] Hicks and Associates, Inc, July 19, 2002. http://epic.org/privacy/profiling/tia/tiasystemdescription.pdf.

25. **Defense Advanced Research Projects Agency.** INFORMATION AWARENESS Solicitation Number: BAA03-23. [Online] Federal Business Opportunities, April 15, 2003. https://www.fbo.gov/index?s=opportunity&mode=form&id=2c05fdcf3acfc3ceb88b1edce3 2ecb7c&tab=core&_cview=1.

26. Information Awareness Proposer Information Pamphlet. [Online] Federal Business Opportunities, April 15, 2003. https://www.fbo.gov/utils/view?id=fb42bb6199c56c7c46530 8f4344826fc.

27. **Hulse, Carl.** THREATS AND RESPONSES: PLANS AND CRITICISMS; Pentagon Prepares A Futures Market On Terror Attacks. [Online] July 29, 2003. http://www.nytimes.com/2003/07/29/us/threats-responses-plans-criticisms-pentagon-prepares-futures-market-terror.html?pagewanted=all&src=pm.

28. **Looney, Robert.** DARPA's Policy Analysis Market for Intelligence: Outside the Box or Off the Wall? [Online] Strategic Insights, September 2003. http://www.au.af.mil/au/awc/awcgate/nps/pam/si_pam.htm.

29. **Scheiber, Noam.** 2003: THE 3rd ANNUAL YEAR IN IDEAS; Futures Markets in Everything. [Online] The New York Times, December 14, 2003. http://www.nytimes.com/2003/12/14/magazine/2003-the-3rd-annual-year-in-ideas-futures-markets-in-everything.html.

30. **Starr, Barbara.** Pentagon folds bets on terror. [Online] CNNMoney, July 29, 2003. http://money.cnn.com/2003/07/29/news/terror_futures/?cnn=yes.

31. **Poindexter, John.** John M. Poindexter Resignation Letter. [Online] The Washington Post, August 12, 2003. http://www.washingtonpost.com/wp-srv/nation/transcripts/poindexterletter.pdf.

32. **U.S. Congress.** Committee Reports. 108th Congress (2003–2004). House Report 108–283. [Online] The Library of Congress, September 2003. http://thomas.loc.gov/cgi-bin/cpquery/?& sid=cp108alJsu&refer=&r_n=hr283.108&db_id=108&item=&&sid=cp108alJsu&r_n=hr 283.108&dbname=cp108&&sel=TOC_309917.

Chapter 4
The Afterlife of Total Information Awareness

> Congress gave me the authority to use necessary force to protect the American people, but it didn't prescribe the tactics.
> —President George W. Bush (January 23, 2006).

> By finally admitting a wrong, a nation does not destroy its integrity but, rather, reinforces the sincerity of its commitment to the Constitution and hence to its people.
> —U.S. Attorney General Dick Thornburgh (October 10, 1990).

> Technology is a two-edged sword for the intelligence community. For instance, with biology, there could be a time in the not distant future when teenagers can design biological components just as they do computer viruses today.
> —ODNI Director of Science and Technology Steven Nixon (2008).

4.1 NSA's Terrorist Surveillance Program

Although the U.S. Congress axed the Information Awareness Office (IAO) and dismantled Total Information Awareness (TIA) in September 2003, TIA did not really cease to exist. Five years later in March 2008, a *Wall Street Journal* article reported that the National Security Agency (NSA) has been building essentially the same system as TIA for its Terrorist Surveillance Program and other U.S. governmental agencies. Wall Street Journal intelligence correspondent Siobhan Gorman wrote [1]:

> According to current and former intelligence officials, the spy agency now monitors huge volumes of records of domestic emails and Internet searches as well as bank transfers, credit-card transactions, travel and telephone records. The NSA receives this so-called "transactional" data from other agencies or private companies, and its sophisticated software programs analyze the various transactions for suspicious patterns.

N. Lee, *Counterterrorism and Cybersecurity*, DOI: 10.1007/978-1-4614-7205-6_4,
© Springer Science+Business Media New York 2013

Two current officials also said the NSA's current combination of programs now largely mirrors the former TIA project. But the NSA offers less privacy protection. TIA developers researched ways to limit the use of the system for broad searches of individuals' data, such as requiring intelligence officers to get leads from other sources first. The NSA effort lacks those controls…

The NSA uses its own high-powered version of social-network analysis to search for possible new patterns and links to terrorism. Former NSA Director Gen. Michael Hayden explained, "The program … is not a driftnet over [U.S. cities such as] Dearborn or Lackawanna or Fremont, grabbing conversations that we then sort out by these alleged keyword searches or data-mining tools or other devices… This is not about intercepting conversations between people in the United States. This is hot pursuit of communications entering or leaving America involving someone we believe is associated with al-Qaeda. … This is focused. It's targeted. It's very carefully done. You shouldn't worry" [2].

In spite of Hayden's assurance, the American Civil Liberties Union (ACLU) issued a statement accusing the NSA of reviving TIA to be an Orwellian domestic spying program [3]:

"Congress shut down TIA because it represented a massive and unjustified governmental intrusion into the personal lives of Americans," said Caroline Fredrickson, Director of the Washington Legislative Office of the ACLU. "Now we find out that the security agencies are pushing ahead with the program anyway, despite that clear congressional prohibition. The program described by current and former intelligence officials in Monday's Wall Street Journal could be modeled on Orwell's Big Brother."

"Year after year, we have warned that our great nation is turning into a surveillance society where our every move is tracked and monitored," said Barry Steinhardt, Director of the ACLU's Technology and Liberty Project. "Now we have before us a program that appears to do that very thing. It brings together numerous programs that we and many others have fought for years, and it confirms what the ACLU has been saying the NSA is up to: mass surveillance of Americans."

The mass surveillance of Americans is a direct violation of the Fourth Amendment to the U.S. Constitution—a Bill of Rights that guards against unreasonable searches and seizures, along with requiring any warrant to be judicially sanctioned and supported by probable cause.

In September 2012, NSA whistleblower William Binney filed a sworn declaration that the agency has installed within the U.S. no fewer than 10 and possibly in excess of 20 intercept centers [4] including the AT&T center on Folsom Street in San Francisco. Binney's testimony supports the revelation from AT&T whistleblower Mark Klein in 2006 [5]. Klein installed inside AT&T's San Francisco switching office a Semantic Traffic Analyzer—an Internet monitoring tool that can reconstruct all of the e-mails sent along with attachments, see what web pages have been clicked on, capture instant messages, and record video streams and VoIP (Voice over Internet Protocol) phone calls [6].

"The danger here is that we fall into a totalitarian state," warned Binney. "This is something the KGB, the Stasi or the Gestapo would have loved to have had" [7].

4.2 President George W. Bush and NSA Warrantless Wiretapping

The National Security Agency (NSA) was created by President Harry Truman in 1952 after World War II to continue the code-breaking work and to prevent another surprise like the attack on Pearl Harbor [8]. The Central Security Service (CSS) was established by President Richard Nixon in 1972 to promote full partnership between NSA and the Service Cryptologic Components of the U.S. Armed Forces [9]. The mission for NSA/CSS is as follows:

> The National Security Agency/Central Security Service (NSA/CSS) leads the U.S. Government in cryptology that encompasses both Signals Intelligence (SIGINT) and Information Assurance (IA) products and services, and enables Computer Network Operations (CNO) in order to gain a decision advantage for the Nation and our allies under all circumstances. The Information Assurance mission confronts the formidable challenge of preventing foreign adversaries from gaining access to sensitive or classified national security information. The Signals Intelligence mission collects, processes, and disseminates intelligence information from foreign signals for intelligence and counterintelligence purposes and to support military operations. This Agency also enables Network Warfare operations to defeat terrorists and their organizations at home and abroad, consistent with U.S. laws and the protection of privacy and civil liberties.

Notwithstanding the constitutional rights of American citizens, the NSA has acted in accordance to a presidential order signed in 2002, shortly after the 9/11 terrorist attacks. *The New York Times* revealed in December 2005 that President George W. Bush "secretly authorized the National Security Agency to eavesdrop on Americans and others inside the United States to search for evidence of terrorist activity without the court-approved warrants ordinarily required for domestic spying" [10]:

> While many details about the program remain secret, officials familiar with it say the NSA eavesdrops without warrants on up to 500 people in the United States at any given time. The list changes as some names are added and others dropped, so the number monitored in this country may have reached into the thousands since the program began, several officials said. Overseas, about 5,000–7,000 people suspected of terrorist ties are monitored at one time, according to those officials.

USA Today also reported that NSA has been secretly collecting the phone call records of tens of millions of Americans, using data provided by AT&T, Verizon, and BellSouth [2]:

> The NSA program reaches into homes and businesses across the nation by amassing information about the calls of ordinary Americans—most of whom aren't suspected of any crime. This program does not involve the NSA listening to or recording conversations. But the spy agency is using the data to analyze calling patterns in an effort to detect terrorist activity. ... It's the largest database ever assembled in the world. The agency's goal is "to create a database of every call ever made" within the nation's borders.

In response to the harsh criticisms from the media, the NSA disclosed some success stories of the domestic eavesdropping program, including foiling a plot by Iyman Faris, a Pakistani American truck driver in Ohio, who wanted to bring down the Brooklyn Bridge in 2002 with blowtorches [11]. The NSA Terrorist Surveillance Program also claimed to have helped thwart the fertilizer bomb attacks on British pubs and train stations in 2004 [12].

In January 2006, President George W. Bush publicly defended NSA's warrantless terrorist surveillance program (aka warrantless wiretapping) that bypassed the 1978 Foreign Intelligence Surveillance Act (FISA). Signed into law by President Jimmy Carter, FISA was introduced by Senator Ted Kennedy to provide judicial and congressional oversight of the government's covert surveillance activities of foreign entities and individuals in the United States without violating the Fourth Amendment to the U.S. Constitution [13].

"If I wanted to break the law, why was I briefing Congress?" Bush told an audience at Kansas State University. "Congress gave me the authority to use necessary force to protect the American people, but it didn't prescribe the tactics" [14]. Bush argued that bypassing the courts fell within presidential power during a time when the country is fighting terrorism. The U.S. Congress sided with Bush and the NSA warrantless wiretapping.

In July 2008, Congress passed the FISA Amendments Act of 2008, which grants retroactive immunity to complicit telecoms and allows eavesdropping in emergencies without court approval for up to seven days [15, 16]. The American Civil Liberties Union (ACLU) filed a lawsuit against FISA Amendments Act of 2008, calling it an "unconstitutional dragnet wiretapping law" [17].

Constitutional rights should be upheld under all circumstances; lest we forget the Japanese American internment during World War II. In October 1990, U.S. Attorney General Dick Thornburgh presented an entire nation's apology to 120,000 Japanese American internees and their descendants. In an emotional reparation ceremony in Washington, Thornburgh said, "By finally admitting a wrong, a nation does not destroy its integrity but, rather, reinforces the sincerity of its commitment to the Constitution and hence to its people" [18].

4.3 Poindexter's Policy Analysis Market

Although Poindexter's controversial Policy Analysis Market (PAM) program was terminated by the U.S. Congress in 2003, a San Diego based private firm that helped develop the software revived PAM in 2004 without involvement from the U.S. government. The new version of PAM would "allow traders to buy and sell contracts on political and economic events in the Middle East, but only on questions that have a positive or neutral slant, such as 'Iraqi oil exports will exceed 2 million barrels a day during the third quarter of 2004" [19].

Economic journalist Noam Scheiber wrote in *The New York Times*, "The beauty of futures markets like PAM is that they're among the most meritocratic institutions ever devised" [20]. He cited the success of Hewlett-Packard's futures market in predicting actual sales [21], the high accuracy of the Iowa Electronic Markets (IEM) in forecasting presidential elections [22], and the impressive performance of the Hollywood Stock Exchange (HSX) in picking Oscar winners [23].

Unlike the controversial PAM that was open to public trading, restricting its access to intelligence officers with credible information can make PAM a powerful decision support tool for the U.S. intelligence community.

4.4 Project Argus: Bio-Surveillance Priming System

After the 9/11 terrorist attacks, Dr. Eric Haseltine left his position as Executive Vice President of R&D at Walt Disney Imagineering to join the National Security Agency (NSA) as Director of Research in 2002. From 2005 to 2007, Haseltine was Associate Director for Science and Technology at the newly established Office of the Director of National Intelligence (ODNI).

In collaboration with Georgetown University researchers, Haseltine and his successor Steven Nixon at ODNI oversaw the development of Argus, a bio-surveillance AI program that monitors foreign news reports and other open sources looking for anything that could provide an early warning of an epidemic, nuclear accident, or environmental catastrophe.

In a statement by Dr. James Wilson of Georgetown University before the U.S. Senate Committee on Homeland Security, Wilson testified in October 2007 [24]:

> Argus is designed to detect and track early indications and warnings of foreign biological events that may represent threats to global health and national security. Argus serves a "tipping function" designed to alert its users to events that may require action. It is not in the business of determining whether, or what type of actions should be taken.
>
> Argus is based on monitoring social disruption through native language reports in electronic local sources around the globe. Argus specifically focuses on taxonomy of direct and indirect types of indications and warnings including:
>
> • Environmental conditions thought to be conducive to support outbreak triggering;
> • Reports of disease outbreaks in humans or animals; and
> • Markers of social disruption such as school closings or infrastructure overloads.
>
> We estimate we are accessing over a million pieces of information daily covering every country in the world which results in producing, on average, 200 reports per day. Using a disease event warning system modeled after NOAA's National Weather Service, we issue Warnings, Watches, and Advisories in accordance with guidelines agreed upon by our research partners in the federal government. On average, we have 15 Advisories, 5 Watches, and 2 Warnings active on our Watchboard at any given time, with 2,200 individual case files of socially disruptive biological events maintained and monitored daily in over 170 countries involving 130 disease entities affecting humans or animals.
>
> Since the program began, we have logged over 30,000 biological events in varying stages of social disruption throughout the world involving pathogens such as H5N1 avian influenza, other influenza strains, Ebola virus, cholera, and other exotic pathogens.
>
> We have discovered the Argus methodology can be made sensitive to events involving nuclear and radiological, chemical, terrorist, political instability, genocide and conflict, crop surveillance, and natural disasters.

Wilson cited the successful tracking of the H3N2 influenza virus spreading from China to Chile, Argentina, Australia, and several other countries. However, Argus was unable to monitor that strain of influenza within the U.S. because domestic monitoring was prohibited

Eric Haseltine said in a 2006 *U.S. News & World Report* interview, "I sleep a little easier at night knowing that Argus is out there" [25]. In 2008, ODNI Director of Science and Technology Steven Nixon told Lawrence Wright of *The New Yorker*, "Technology is a two-edged sword for the intelligence community. For instance, with biology, there could be a time in the not distant future when teenagers can design biological components just as they do computer viruses today. That's why I think intelligence is as critical now as at any time in our nation's history" [26].

4.5 President Barack Obama's Big Data R&D Initiative

At the Defense Advanced Research Projects Agency (DARPA), the dismantled Information Awareness Office (IAO) has been replaced by the Information Innovation Office (I2O). I2O has been carrying on research projects that were not shut down by the U.S. Congress since the termination of Total Information Awareness (TIA) [27]. They include Bio-Event Advanced Leading Indicator Recognition Technology, Rapid Analytic Wargaming, Wargaming the Asymmetric Environment, and Automated Speech and Text Exploitation in Multiple Languages (including Babylon and Symphony) [28].

Total Information Awareness requires efficient and effective data mining. In March 2012, the Obama administration announced more than $200 million in funding for the "Big Data Research and Development Initiative" [29]. The first wave of agency commitments includes National Science Foundation (NSF), National Institutes of Health (NIH), Department of Energy (DOE), U.S. Geological Survey, and Department of Defense (including DARPA) [30].

Among the funded DARPA programs is Anomaly Detection at Multiple Scales (ADAMS), one of several key technologies that were directly applicable to Total Information Awareness [31].

"In the same way that past Federal investments in information-technology R&D led to dramatic advances in supercomputing and the creation of the Internet, the initiative we are launching today promises to transform our ability to use Big Data for scientific discovery, environmental and biomedical research, education, and national security," said Dr. John P. Holdren, Assistant to the President and Director of the White House Office of Science and Technology Policy [30].

4.6 CIA's In-Q-Tel Funded Palantir Technologies

Palantir Technologies was founded in 2004 by Peter Thiel (PayPal cofounder), Alex Karp, Joe Lonsdale, Stephen Cohen, and Nathan Gettings. With early investments from Thiel and the Central Intelligence Agency (CIA) venture arm In-Q-Tel, Palantir develops software applications for integrating, visualizing and analyzing big data that is structured, unstructured, relational, temporal, and geospatial [32].

"Using Palantir technology," *Bloomberg Businessweek* reported in November 2011, "the FBI can now instantly compile thorough dossiers on U.S. citizens, tying together surveillance video outside a drugstore with credit-card transactions, cell-phone call records, e-mails, airplane travel records, and Web search information" [33].

The National Center for Missing and Exploited Children (NCMEC) has also used Palantir software to solve child abuse and abduction cases. Ernie Allen, CEO of NCMEC, praised Palantir for "the ability to do the kind of link-and-pattern analysis we need to build cases, identify perpetrators, and rescue children" [33].

Built upon PayPal's fraud detection algorithms, Palantir excels in discovering connections between seemingly unrelated incidents as well as the people involved. Besides counterterrorism and law enforcement, Palantir applications have been deployed in banks, hospitals, law firms, insurance companies, pharmaceuticals, and other organizations [34].

However, Palantir's involvement in a convoluted plot to bring down WikiLeaks in 2011 raised some eyebrows [35]. Furthermore, Palantir's senior counsel Bryan Cunningham was former National Security Council legal adviser who supported President George W. Bush's authorization of the NSA warrantless wiretapping [36].

Peter Thiel told *Bloomberg Businessweek* in defense of Palantir, "We cannot afford to have another 9/11 event in the U.S. or anything bigger than that. That day opened the doors to all sorts of crazy abuses and draconian policies. The best way to avoid such scenarios in the future would be to provide the government the most cutting-edge technology possible and build in policing systems to make sure investigators use it lawfully" [33].

Nonetheless, Christopher Soghoian, principal technologist at the American Civil Liberties Union (ACLU) voiced his concerns while he was a graduate fellow at Indiana University: "I don't think Palantir the firm is evil. I think their clients could be using it for evil things." In regard to Palantir's built-in privacy protection features, Soghoian said, "If you don't think the NSA can disable the piece of auditing functionality, you have to be kidding me. They can do whatever they want, so it's ridiculous to assume that this audit trail is sufficient" [33].

4.7 Microsoft and NYPD's Domain Awareness System

In April 2009, the New York Police Department (NYPD) has developed a real-time networked Domain Awareness System (DAS) to detect, deter, and prevent potential terrorist activities in New York City [37]. As part of the counterterrorism program of the NYPD's Counterterrorism Bureau, the deployed DAS technology includes closed-circuit televisions (CCTVs), License Plate Readers (LPRs), and other domain awareness devices.

The CCTVs are operated by NYPD as well as partnering companies and government agencies that provide feeds from their proprietary CCTVs into the Lower Manhattan Security Coordination Center. License plate data are collected by fixed or mobile LPR devices. Other domain awareness devices gather environmental data and detect hazards.

In August 2012, Microsoft and NYDP jointly announced bringing the DAS technology to law enforcement agencies around the world. According to retired U.S. Army Lt. Gen. Mike McDuffie, DAS "aggregates and analyzes public safety data in real time and combines artificial intelligence analytics with video from around a jurisdiction to identify potential threats and protect critical infrastructure" [38].

4.8 NSA's $2-Billion-Dollar Data-Mining and Spy Center

On September 6, 2009, an email was sent by a suspected al-Qaeda member "Ahmad" from Pakistan to a previously unknown man in Denver, Colorado. It was instantly logged by the NSA computers in Fort Meade, Maryland. The Denver-based emailer, Najibullah Zazi, was later convicted of a suicide bombing plot against the New York subway [39].

"Forty years ago there were 5,000 stand-alone computers, no fax machines and not one cellular phone," said Former NSA director Gen. Michael Hayden. "Today there are over 180 million computers—most of them networked. There are roughly 14 million fax machines and 40 million cellphones, and those numbers continue to grow" [40]. With more than a dozen listening posts around the world, the NSA intercepts about two million phone calls, e-mail messages, faxes and other types of communications every hour.

The skyrocketing volume of information is to be stored and handled at the nation's largest data-mining center at Camp Williams National Guard Training Site in Bluffdale, Utah. The new construction was given the "green light" when President Barack Obama signed the 2009 Supplemental-War Funding Bill [41]. The NSA broke ground on the facility in January 2011. The 1–1.5-million-square-foot, $1.5–$2-billion-dollar spy center is slated for completion in September 2013.

Figure 4.1 shows the conceptual site plan for the NSA data center with eight main areas:

1. Visitor control center—A $9.7 million facility for ensuring that only authorized personnel gain access.
2. Administration—A 900,000-square-foot space for technical support and administrative personnel.
3. Data halls—Four 25,000-square-foot facilities house rows of servers.
4. Backup generators and fuel tanks—They can power the center for three days in an emergency.
5. Water storage and pumping—They are able to pump 1.7 million gallons of liquid per day.

Fig. 4.1 NSA's new data center conceptual site plan (Courtesy of the U.S. Army Corps of Engineers)

6. Chiller plant—About 60,000 tons of cooling equipment keep servers from overheating.
7. Power substation—An electrical substation to meet the center's estimated 65 MW demand.
8. Security—Video surveillance, intrusion detection, and other antiterrorism protection at a cost of $10 million.

James Bamford, investigative journalist and former Navy intelligence analyst, wrote in a March 2012 issue of the *Wired* magazine, "Once it's operational, the Utah Data Center will become, in effect, the NSA's cloud. The center will be fed data collected by the agency's eavesdropping satellites, overseas listening posts, and secret monitoring rooms in telecom facilities throughout the US. All that data will then be accessible to the NSA's code breakers, data-miners, China analysts, counterterrorism specialists, and others working at its Fort Meade headquarters and around the world" [42].

NSA deputy director John Chris Inglis said, "It's a state-of-the-art facility designed to support the intelligence community in its mission to, in turn, enable and protect the nation's cybersecurity." But an unnamed senior intelligence officer told Bamford, "This is more than just a data center. It is also critical for breaking codes. … Everybody's a target; everybody with communication is a target" [42].

Investigative journalist Shane Harris compared Poindexter's TIA with the NSA data-mining project in his 2012 article "Giving In to the Surveillance State" published in *The New York Times* [43]:

Today, this global surveillance system continues to grow. It now collects so much digital detritus—e-mails, calls, text messages, cellphone location data and a catalog of computer viruses—that the NSA is building a 1-million-square-foot facility in the Utah desert to store and process it.

What's missing, however, is a reliable way of keeping track of who sees what, and who watches whom. After TIA was officially shut down in 2003, the NSA adopted many of Mr. Poindexter's ideas except for two: an application that would "anonymize" data, so that information could be linked to a person only through a court order; and a set of audit logs, which would keep track of whether innocent Americans' communications were getting caught in a digital net.

References

1. **Gorman, Siobhan.** NSA's Domestic Spying Grows As Agency Sweeps Up Data. [Online] The Wall Street Journal, March 10, 2008. http://online.wsj.com/article/SB120511973377523845.html.
2. **Cauley, Leslie.** NSA has massive database of Americans' phone calls. [Online] USA Today, May 11, 2006. http://www.usatoday.com/news/washington/2006-05-10-nsa_x.htm.
3. **American Civil Liberties Union.** Stunning New Report on Domestic NSA Dragnet Spying Confirms ACLU Surveillance Warnings. [Online] American Civil Liberties Union, March 12, 2008. http://www.aclu.org/technology-and-liberty/stunning-new-report-domestic-nsa-dragnet-spying-confirms-aclu-surveillance-wa.
4. **CASE NO. CV-08-04373-JSW.** Declaration Of William E. Binney In Support Of Plaintiffs' Motion For Partial Summary Judgment Rejecting The Government Defendants' State Secret Defense. [Online] United States District Court for the Northern District of California, September 28, 2012. http://info.publicintelligence.net/NSA-WilliamBinneyDeclaration.pdf.
5. **C-06-0672-VRW.** Declaration of Mark Klein in Support of Plainteiffs' Motion for Preliminary Injunction. [Online] United States District Court Northern District of California, June 8, 2006. https://www.eff.org/files/filenode/att/SER_klein_decl.pdf.
6. **Poe, Robert.** The Ultimate Net Monitoring Tool. [Online] Wired, May 17, 2006. http://www.wired.com/science/discoveries/news/2006/05/70914.
7. **Kelley, Michael.** NSA Whistleblower Details How The NSA Has Spied On US Citizens Since 9/11. [Online] Business Insider, August 24, 2012. http://www.businessinsider.com/nsa-whistleblower-william-binney-explains-nsa-surveillance-2012-8.
8. **National Security Agency.** Our History. [Online] National Security Agency, January 15, 2009. http://www.nsa.gov/public_info/speeches_testimonies/nsa_videos/history_of_nsa.shtml.
9. Central Security Service (CSS). [Online] National Security Agency, November 21, 2012. http://www.nsa.gov/about/central_security_service/index.shtml.
10. **Risen, James and Lichtblau, Eric.** Bush Lets U.S. Spy on Callers Without Courts. [Online] The New York Times, December 16, 2005. http://www.nytimes.com/2005/12/16/politics/16program.html?pagewanted=all.
11. **Department of Justice.** Iyman Faris Sentenced for Providing Material Support to Al Qaeda. [Online] U.S. Department of Justice, October 28, 2003. http://www.justice.gov/opa/pr/2003/October/03_crm_589.htm.
12. **Summers, Chris and Casciani, Dominic.** Fertiliser bomb plot: The story. [Online] BBC News, April 30, 2007. http://news.bbc.co.uk/2/hi/uk_news/6153884.stm.
13. **epic.org.** Foreign Intelligence Surveillance Act (FISA). [Online] Electronic Privacy Information Center. http://epic.org/privacy/terrorism/fisa/.
14. **Sanger, David E. and O'Neil, John.** White House Begins New Effort to Defend Surveillance Program. [Online] The New York Times, January 23, 2006. http://www.nytimes.com/2006/01/23/politics/23cnd-wiretap.html?pagewanted=all.
15. **110th Congress.** Foreign Intelligence Surveillance Act Of 1978 Amendments Act Of 2008. [Online] U.S. Senate Select Committee on Intelligence, July 10, 2008. http://www.intelligence.senate.gov/laws/pl110261.pdf.
16. **Lichtblau, Eric.** Senate Approves Bill to Broaden Wiretap Powers. [Online] The New York Times, July 10, 2008. http://www.nytimes.com/2008/07/10/washington/10fisa.html.

17. **ACLU.** ACLU Sues Over Unconstitutional Dragnet Wiretapping Law. [Online] American Civil Liberties Union, July 10, 2008. http://www.aclu.org/national-security/aclu-sues-over-unconstitutional-dragnet-wiretapping-law.
18. **Ostrow, Ronald J.** First 9 Japanese WWII Internees Get Reparations. [Online] Los Angeles Times, October 10, 1990. http://articles.latimes.com/1990-10-10/news/mn-1961_1_japanese-wwii-internees.
19. **Gongloff, Mark.** Middle East futures market returns. Private firm will restart Pentagon project, but without contracts for violence, in 2004. [Online] CNNMoney, November 18, 2003. http://money.cnn.com/2003/11/17/news/terror_futures/index.htm.
20. **Scheiber, Noam.** 2003: THE 3rd ANNUAL YEAR IN IDEAS; Futures Markets in Everything. [Online] The New York Times, December 14, 2003. http://www.nytimes.com/2003/12/14/magazine/2003-the-3rd-annual-year-in-ideas-futures-markets-in-everything.html.
21. **Chen, Kay-Yut and Plott, Charles R.** Information Aggregation Mechanisms: Concept, Design and Implementation for a Sales Forecasting Problem. [Online] California Institute of Technology, March 2002. www.hpl.hp.com/personal/Kay-Yut_Chen/paper/ms020408.pdf.
22. **The University of Iowa.** Iowa Electronic Markets. [Online] University of Iowa Henry B. Tippie College of Business. http://tippie.uiowa.edu/iem/.
23. **HSX.** Hollywood Stock Exchange: The Entertainment Market. [Online] http://www.hsx.com/.
24. **Wilson, James M.** Statement by James M. Wilson V, MD. [Online] U.S. Senate Committee on Homeland Security and Governmental Affairs, October 4, 2007. http://www.hsgac.senate.gov//imo/media/doc/WilsonTestimony.pdf.
25. **U.S. News & World Report.** Q&A: DNI Chief Scientist Eric Haseltine. [Online] U.S. News & World Report, November 3, 2006. http://www.usnews.com/usnews/news/articles/061103/3qahaseltine_6.htm.
26. **Wright, Lawrence.** The Spymaster. Can Mike McConnell fix America's intelligence community? [Online] The New Yorker, January 21, 2008. http://www.newyorker.com/reporting/2008/01/21/080121fa_fact_wright?currentPage=all.
27. **DARPA.** Information Innovation Office. [Online] Defense Advanced Research Projects Agency. http://www.darpa.mil/Our_Work/I2O/.
28. **U.S. Congress.** Committee Reports. 108th Congress (2003-2004). House Report 108-283. [Online] The Library of Congress, September 2003. http://thomas.loc.gov/cgi-bin/cpquery/?&sid=cp108alJsu&refer=&r_n=hr283.108&db_id=108&item=&&sid=cp108alJsu&r_n=hr283.108&dbname=cp108&&sel=TOC_309917.
29. **Kalil, Tom.** Big Data is a Big Deal. [Online] The White House, March 29, 2012. http://www.whitehouse.gov/blog/2012/03/29/big-data-big-deal.
30. **Office of Science and Technology Policy.** Obama Administration Unveils "Big Data" Initiative: Announces $200 Million In New R&D Investments. [Online] Executive Office of the President, March 29, 2012. http://www.whitehouse.gov/sites/default/files/microsites/ostp/big_data_press_release.pdf.
31. **Executive Office of the President.** Big Data Across the Federal Government. [Online] The White House, March 29, 2012. http://www.whitehouse.gov/sites/default/files/microsites/ostp/big_data_fact_sheet_final_1.pdf.
32. **Gorman, Siobhan.** How Team of Geeks Cracked Spy Trade. [Online] The Wall Street Journal, September 4, 2009. http://online.wsj.com/article/SB125200842406984303.html.
33. **Vance, Ashlee and Stone, Brad.** Palantir, the War on Terror's Secret Weapon. [Online] Bloomberg Businessweek, November 22, 2011. http://www.businessweek.com/magazine/palantir-the-vanguard-of-cyberterror-security-11222011.html.
34. **Palantir Technologies.** Industries & Solutions. [Online] Palantir Technologies. [Cited: November 14, 2012.] http://www.palantir.com/solutions/.
35. **Anderson, Nate.** Spy Games: Inside the Convoluted Plot to Bring Down WikiLeaks. [Online] Wired, February 14, 2011. http://www.wired.com/threatlevel/2011/02/spy/.
36. **Angle, Jim and Herridge, Catherine.** Debate Rages Over Legality of NSA Wiretap Program. [Online] Fox News, December 21, 2005. http://www.foxnews.com/story/0,2933,179323,00.html.

37. **New York City Police Department.** Public Security Privacy Guidelines. [Online] New York City Police Department, April 2, 2009. http://www.nyc.gov/html/nypd/downloads/pdf/crime_prevention/public_security_privacy_guidelines.pdf.

38. **McDuffie, Mike.** Microsoft and NYPD Announce Partnership Providing Real-Time Counterterrorism Solution Globally. [Online] Microsoft, August 8, 2012. http://www.microsoft.com/government/en-us/state/brightside/Pages/details.aspx?Microsoft-and-NYPD-Announce-Partnership-Providing-Real-Time-Counterterrorism-Solution-Globally&blogid=697.

39. **Cruickshank, Paul.** Inside the plot to devastate New York. [Online] CNN, May 2, 2012. http://security.blogs.cnn.com/2012/05/02/time-line-for-a-terror-plot/.

40. **Bamford, James.** War of Secrets; Eyes in the Sky, Ears to the Wall, and Still Wanting. [Online] The New York Times, September 8, 2002. http://www.nytimes.com/2002/09/08/weekinreview/war-of-secrets-eyes-in-the-sky-ears-to-the-wall-and-still-wanting.html?pagewantcd=all.

41. **Draughn, Katisha.** Federal partners break ground on $1.5 billion center. [Online] US Army Corps of Engineers Baltimore District, February 10, 2011. http://www.nab.usace.army.mil/News/20110210_Federal%20partners%20break%20ground%20on%20$1.5%20billion%20center.htm.

42. **Bamford, James.** The NSA Is Building the Country's Biggest Spy Center (Watch What You Say). [Online] Wired, March 15, 2012. http://www.wired.com/threatlevel/2012/03/ff_nsadatacenter/all/.

43. **Harris, Shane.** Giving Into the Surveillance State. [Online] The New York Times, August 22, 2012. http://www.nytimes.com/2012/08/23/opinion/whos-watching-the-nsa-watchers.html.

Chapter 5
Artificial Intelligence and Data Mining

A lot of cutting edge AI has filtered into general applications, often without being called AI because once something becomes useful enough and common enough it's not labeled AI anymore.
—Nick Bostrom.
Oxford University Future of Humanity Institute (2006).

Whenever an AI research project made a useful new discovery, that product usually quickly spun off to form a new scientific or commercial specialty with its own distinctive name.
—Professor Marvin Minsky.
MIT Artificial Intelligence Laboratory (2009).

HAL's not the focus; the focus is on the computer on 'Star Trek'.
—David Ferrucci.
IBM Thomas J. Watson Research Center (2011).

5.1 Artificial Intelligence: From Hollywood to the Real World

In 1955, American computer scientist and cognitive scientist John McCarthy coined the term "artificial intelligence" (AI). He defined AI as "the science and engineering of making intelligent machines, especially intelligent computer programs" [1]. In the 2001 film *A.I.: Artificial Intelligence*, Steven Spielberg tells the story of a highly advanced robotic boy who longs to become real so that he can regain the love his human mother [2]. In 2004, Will Smith starred in the lead role of *I, Robot*—a film based loosely on Isaac Asimov's short-story collection of the same name [3]. Although the Hollywood movies are quite far-fetched, AI hit the spotlight on primetime television over three nights in February 2011 when the IBM Watson computer won on "Jeopardy!" against two human champions and took home a $1 million prize [4].

N. Lee, *Counterterrorism and Cybersecurity*, DOI: 10.1007/978-1-4614-7205-6_5,
© Springer Science+Business Media New York 2013

Watson, named after IBM founder Thomas J. Watson, has the ability of encyclopedic recall and natural language understanding. "People ask me if this is HAL," said David Ferrucci, lead developer of Watson, referring to the Heuristically programmed ALgorithmic (HAL) computer in *2001: A Space Odyssey* by Stanley Kubrick and Arthur C. Clarke. "HAL's not the focus; the focus is on the computer on 'Star Trek,' where you have this intelligent information seek dialogue, where you can ask follow-up questions and the computer can look at all the evidence and tries to ask follow-up questions. That's very cool" [5].

Watson was inspired by the Deep Blue project at IBM. Back in May 1997, the IBM Deep Blue computer beat the world chess champion Garry Kasparov after a six-game match, marking the first time in history that a computer had ever defeated a world champion in a match play [6]. Since then, computers have become much faster and software more sophisticated. In October 2012, the U.S. Department of Energy unveiled the Titan supercomputer capable of 20 petaflops—20 thousand trillion (20,000,000,000,000,000) floating point operations per second [7].

Although our desktop computers are no match for the Titan, AI software has entered mainstream consumer products. Apple's intelligent personal assistant Siri on iPhone, iPad, and iPod is the epitome of AI in everyday life. Siri uses voice recognition and information from the user's contacts, music library, calendars, and reminders to better understand what the user says [8]. The software application is an offshoot of SRI International's Cognitive Assistant that Learns and Organizes (CALO) project funded by the Defense Advanced Research Projects Agency (DARPA) under its Perceptive Assistant that Learns (PAL) program [9, 10]. Apple acquired Siri in April 2010, integrated it into iOS, and the rest is history [11].

In addition to smartphones, domain-specific AI software applications have been embedded into newer automobiles, interactive toys, home appliances, medical equipment, and many electronic devices.

We do not often hear about AI in the real world, because as MIT Professor Marvin Minsky explained, "AI research has made enormous progress in only a few decades, and because of that rapidity, the field has acquired a somewhat shady reputation! This paradox resulted from the fact that whenever an AI research project made a useful new discovery, that product usually quickly spun off to form a new scientific or commercial specialty with its own distinctive name" [12].

Professor Nick Bostrom, director of the Future of Humanity Institute at Oxford University, told CNN in a 2006 interview, "A lot of cutting edge AI has filtered into general applications, often without being called AI because once something becomes useful enough and common enough it's not labeled AI anymore" [13].

5.2 Intelligent CCTV Cameras

Artificial intelligence (AI) is increasingly used in the processing of collected data from physical surveillance. There are approximately 30 million closed-circuit television (CCTV) cameras in the world capturing 250 billion hours of raw footage annually [14]. It is time-prohibitive to manually process that much data in search of clues that will solve a crime. The effort would be like looking for a needle in a haystack. In addition, while CCTV cameras help deter crimes, they are less effective than an eyewitness at the scene who can alert the police.

AI software empowers CCTV surveillance by automatic detection of visual and audio clues to spot anything out of the ordinary such as violent crimes, vandalism, and terrorism. Not only can AI software scan the recorded footages at high speed, it can also do real-time analysis at the scene much like a human eyewitness.

Prof. David Brown at the University of Portsmouth described the intelligent CCTV cameras that can see and hear, and that can alert law enforcement authorities to crimes in progress [15]:

> We have already developed visual recognition software, but the next stage is to develop audio recognition software to listen for particular sounds. We can teach the cameras to listen out for things like a swear word being shouted in an aggressive way, or for other words which might signify a crime taking place. The camera will be able to swivel to the direction of the sound at the same speed someone turns their head when they hear a scream, or about 300 ms. People monitoring CCTV images have banks of screens in front of them, and this system helps them by alerting them to something the system has spotted. The person looking at the screen can then quickly identify if it is a crime taking place, or whether the camera has simply picked up on something innocent, like a child screaming, and act on it accordingly. The system would not be sensitive enough to record individual conversations. We are just looking for certain trigger sounds and visual anomalies.

Equipped with facial recognition capabilities, AI-enhanced CCTV systems have been used in cities like Chicago and London [16]. Although the systems can be susceptible to false alarms, a human counterpart can make a judgment call upon receiving an alert. Artificial intelligence (AI) assisting human decision making is a form of cognitive augmentation or intelligence amplification (IA).

In 2009, Microsoft teamed up with New York Police Department (NYPD) in developing Domain Awareness System (DAS)—a real-time networked counterterrorism system to detect, deter, and prevent potential terrorist activities in New York City [17]. Companies partnering with NYPD provide feeds from their proprietary CCTVs into the Lower Manhattan Security Coordination Center.

In 2012, Microsoft and NYDP announced that DAS will "feature the use of artificial intelligence (AI) capabilities to analyze video, public safety data and other situational awareness information in real time to proactively identify potential terrorist threats and protect critical infrastructure" [18].

5.3 Data Mining in the Age of Big Data

Slated for completion in September 2013, the nation's largest data-mining center at Camp Williams National Guard Training Site in Bluffdale, Utah is to handle the skyrocketing volume of information collected by the National Security Agency (NSA) [19]. With more than a dozen listening posts around the world, the NSA intercepts about two million phone calls, e-mail messages, faxes and other types of communications every hour [20].

Analysts at the NSA, Central Intelligence Agency (CIA), and Federal Bureau of Investigation (FBI) have been knee-deep in a mountain of collected data from physical surveillance and open source information, looking for useful patterns.

A 2001 Congressional report disclosed that the NSA was faced with "profound needle-in-the-haystack challenges." *The New York Times* revealed in 2002 that there were 200 million pieces of intelligence in a regular workday, and less than one percent of it was ever decoded, translated, or processed [21]. Recognizing the importance of data mining, the Obama administration in March 2012 announced more than $200 million in funding for the "Big Data Research and Development Initiative" [22].

Association for Computing Machinery (ACM) Special Interest Group on Knowledge Discovery and Data Mining (SIGKDD) defines data mining as "an interdisciplinary field at the intersection of artificial intelligence, machine learning, statistics, and database systems" [23]. The goal of data mining is to extract knowledge from the available data by capturing this knowledge in a human-understandable structure. The discovery of structure in big data involves:

1. Database, data management, and data warehouse structure.
2. Data preprocessing, transformations, and dimensionality.
3. Choice of model, valid approximations, and statistical inference considerations.
4. Interestingness metrics and choice of algorithms.
5. Algorithmic complexity and scalability considerations.
6. Post-processing of discovered structure.
7. Visualization and understandability.
8. Maintenance, updates, and model life cycle considerations.

Voluminous amounts of structured and unstructured data residing in a vast number of heterogeneous databases present a real challenge to data mining. For example:

1. Despite the establishment of the Terrorist Screening Center (TSC) in 2003, U.S. agencies handling the terrorist watch lists have continued to "work from at least 12 different, sometimes incompatible, often uncoordinated and technologically archaic databases" [24].
2. For spy agencies like the CIA and military intelligence organizations, *The Wall Street Journal* revealed in 2009 that there are hundreds of databases used by each and most of them are not linked up [25].
3. In July 2010, Michael T. Flynn, Deputy Chief of Staff of Intelligence in Afghanistan, wrote a memorandum citing the urgent need for a new system to analyze the vast amounts of intelligence being collected. Flynn wrote, "US intelligence analysts in Afghanistan have several tools available to access the ever-increasing amount of intelligence and battlefield information residing in a myriad of databases. These tools provide access to the information, some more readily than others, but provide little in the way of improved analytical support" [26].

5.4 Knowledge Representation, Acquisition, and Inference

As a co-pioneer of artificial intelligence applications in counterterrorism, I helped develop a natural language parser and machine learning program to digest news and articles in search of potential terrorist threats around the globe. Those were the early days of data mining at the Institute for Defense Analyses (IDA) in 1984.

Employing psychology and cognitive science, my prototype system thinks like a human in constructing small-scale models of reality that it uses to anticipate events [27]. The knowledge representation and data structures were based on "A Framework for Representing Knowledge" by MIT Artificial Intelligence Lab co-founder Marvin Minsky [28]. Automated data analysis applies models to data in order to predict behavior, assess risk, and determine associations. The models can be based on patterns obtained from data mining or based on subjects under surveillance [29].

Artificial intelligence can assist human analysts in data mining by more efficiently organizing the information, detecting missing pieces of data, and making inferences that may otherwise be overlooked by the human eyes. The three fundamental artificial intelligence (AI) techniques are:

1. Knowledge representation

Data mining seeks to discover useful patterns which can take various forms of knowledge presentation. Some of the knowledge representation techniques are heuristic question answering, neural networks, theorem proving, and expert systems. Programming languages for knowledge representation include s-expression-based Lisp (List processing) [30], rule-based Prolog (Programming in logic) [31], and frame-based KL-ONE [32].

Which kind of knowledge representation is best? Marvin Minsky offered his answer: "To solve really hard problems, we'll have to use several different representations. This is because each particular kind of data structure has its own virtues and deficiencies, and none by itself seems adequate for all the different functions involved with what we call 'common sense.' Each has domains of competence and efficiency, so that one may work where another fails. Furthermore, if we rely only on any single 'unified' scheme, then we'll have no way to recover from failure" [33].

2. Knowledge acquisition

Knowledge acquisition from databases involves both database technologies and machine learning techniques. Databases may be centralized, distributed, hierarchical, relational, object-oriented, spatial, temporal, real-time, unstructured, or any combinations of these [34]. An email, for instance, contains an unstructured message along with IP-based geolocation, timestamp, and email addresses of the sender and recipient.

Machine learning can speed up the processing of the million pieces of daily intelligence by automatically parsing, classifying, and reorganizing data [35]. AI not only can streamline the workflow but it can also detect missing pieces of information that need to be acquired in order to form a complete picture.

3.   Knowledge inference

Expert systems can learn from human analysts in making inferences based on the patterns discovered from data mining. The inference engine within an expert system can use propositional, predicate, modal, deontic, temporal, or fuzzy logic to conduct forward chaining, backward chaining, abduction, and reasoning under uncertainty. In the medical domain, expert systems such as MYCIN have shown to outperform doctors in diagnosing diseases [36]. In experimental design, expert systems can study a large number of variables simultaneously and analyze the resulting data using variance decomposition methods [37].

While expert systems are proven to be superior in some cases, they are not meant to replace human analysts but rather they are invaluable tools to supplement human intelligence. For example, in the 2005 computer-assisted PAL/CSS Freestyle Chess Tournament, two amateur chess players used three computers for analysis and won the tournament by defeating all the other teams including grandmasters who were 1,000 Elo points stronger and equipped with more powerful computers [38]. The amateurs turned out to be better at human–computer symbiosis than the grandmasters. Human–computer symbiosis is the idea that technology should be designed in a way that amplifies human intelligence instead of attempting to replace it.

5.5 Dynamic Mental Models

I left the Institute for Defense Analyses for Bell Laboratories in 1985 to further my research on artificial intelligence and expert systems. At Bell Labs, I conceived Dynamic Mental Models (DM^2) as a general algorithm that combines analytical models and experiential knowledge in diagnostic problem solving, regardless of the problem domains [39]. The algorithm mimics a human expert in formulating and using an internal, cognitive representation of a physical system during the process of diagnosis. This internal representation, known as a mental model, originates from an analytical model but it changes dynamically to various levels of abstraction that are most appropriate for efficient diagnosis. An analytical model is represented as structure and behavior, whereas experiential knowledge is expressed in terms of pattern-recognition, topological clustering, topological pruning, and recommendation rules.

Realizing that rules alone are insufficient to make an expert system as smart as a competent human being, some AI researchers advocate reasoning from first principles which are the structure (components and their interconnections) and the behavior (input and output characteristics) of a given system [40, 41]. Such an approach is known as model-based reasoning, which promises some progress towards achieving the goals of application versatility, program understandability, knowledge base extensibility, ease of maintenance, and capability of dealing with novel situations [42]. However, model-based systems are often computationally more expensive than their rule-based counterparts. The situation is worse for applications where certain required computations, such as functional inversion, are practically impossible. A major reason for this difficulty is that the model-based approach relies heavily on the analytical models of a given physical system and undermines the power of experiential knowledge that human experts possess.

Both analytical models and experiential knowledge are essential in building powerful expert systems. To justify this statement, let us take another look at the conventional rule-based approach and ask ourselves the question: "What's in a rule?" A rule such as "if the body temperature elevates abnormally, then the subject has a fever" is a very natural way of expressing a diagnostic decision. However, the rule does little to explain the reasons behind the decision. To justify the decision, one begins to think about various entities and relationships implied by the rule. It is obvious that there must be a human body which has a central temperature. What a rule embodies, therefore, is an implicit model of a domain and the explicit experience of a human expert in diagnosing a problem. In other words, a rule is an end-product that results from compiling domain-specific facts and personal experience. Capturing all the rules (if at all possible) in an expert system is insufficient to make the system as smart as a human expert. Unlike a novice, an expert understands why a rule is a rule and is therefore able to change the rule whenever necessary. Capturing this ability is a challenging problem for AI. To begin with, expert systems have to "de-compile" the rules to make explicit both the domain models and the experiential knowledge.

In 1987, the DM^2 algorithm was implemented and tested on a real-world expert system prototype for telecommunication networks maintenance at AT&T. The application demonstrated that the dynamic mental model approach promotes system robustness, program correctness, software reuse, and ease of knowledge base modification and maintenance [39]. In 1989, the U.S. Army Research Office studied DM^2 for use in diagnostic support of complex modern weapons systems with promising results [43]. In 2004, the U.S. Naval Research Laboratory applied dynamic mental models to meteorological forecasting [44].

5.6 Modeling Human Problem Solving

Research in cognitive psychology suggests that human beings employ the so-called mental models to understand knowledge about the physical world [45]. Mental models differ significantly from analytical models.

An analytical model of a physical system is a result of some engineering design or scientific investigation which is often documented in books, manuals, and written reports. Such a model is an accurate, consistent, and objective representation of a physical system.

On the contrary, a mental model is naturally evolving. Experiments have indicated that a novice reasons about a physical system by first creating a crude, buggy initial mental model of the system and then successively refining it to more elaborated models [46].

Experts are generally much better problem solvers than novices. The reason is that experts know the correct and powerful analytical models for a given physical system, and they also possess rich experience in reasoning with those models. As a result, they are able to formulate much better mental models in terms of accuracy, consistency, and objectivity.

A correct and powerful analytical model needs not to be complete to the lowest possible level of details. Taking an example from the domain of very large-scale integrated circuits (VLSI), a good analytical model describes the circuit at the logic-gate level, not at the level of transistors and resistors. Taking another example in the domain of internal medicine, a given problem may require a "fuzzy" analytical model of a human body that describes the functions and interrelations of bodily organs, whereas another problem may require the inclusion of sensory receptors in the model.

Human experts, upon analyzing the given problem at hand, know what kinds of analytical models are most appropriate. The experts formulate their own mental models of a physical system based on its analytical model as well as personal experience such as undocumented information about the failure rates of certain devices. The experts continue to modify their mental models until they successfully solve the given problem.

Modeling the way human experts solve problems helps improve the performance of AI software programs. In addition, it helps to uncover subtle erroneous decisions and beliefs that human experts might take on in some situations.

5.7 Structural Topology and Behavioral Causality

The mental model of a physical system is defined as the structural topology and behavioral causality of that system as perceived by a human mind.

Structural topology refers to the components and their interconnections within a given system. A topology provides information about not only the existence of certain constituents in a system but also the possible flow of data within a system. In VLSI, for example, a typical structural topology is a circuit diagram showing some integrated circuit chips and their interconnections by which logical truth values propagate from one chip to another.

Behavioral causality refers to the cause-effect relationships between the inputs and outputs of a given component in a system. The behavior of an entire system is the "sum total" of all individual behaviors of its components. A behavior can be represented as a rule, an equation, or a procedure. In electronic devices, for instance, an ideal transformer has a behavioral description of the form: $Z(p) = (N(p)/N(s))^2 * Z(s)$ where Z is the impedance and N is the number of turns on the primary coil p and the secondary coil s. Besides inputs and outputs, a behavior can describe discrete physical states of a device such as "open" and "close" of an on/off switch. The behavioral description of a device is usually limited to its intended function. In other words, we are normally interested in a subset of behaviors that is relevant to the diagnostic problem at hand. Such a subset is called a "relevant facet of behavior."

5.8 Component Clustering and Decoupling

"Component clustering" and "component decoupling" describe the mechanisms that human experts use to modify their mental models. Component clustering refers to the agglomeration of physically adjacent components to form a single composite element whose behavior is the sum total of all its constituents. On the contrary, component decoupling means separating apart adjacent components from a single composite element.

By clustering physically adjacent components we increase the level of abstraction on various parts of a mental model and thereby reduce its complexity. Clustering is a sensible thing to do when we believe that the faults are not located at any of the components that are to be agglomerated. Should our belief be proven incorrect, we would modify the mental model by decoupling the clusters. It is permissible to have multi-layered clusters, that is, clusters within clusters.

By decoupling a composite element, we refine various parts of a mental model and thereby increase its intricacy. Decoupling is inevitable when the faults are believed to reside in one or more of the clustered components.

5.9 Analytical Models and Experiential Knowledge

Analytical models provide the basis for formulating useful mental models for diagnosis. Experiential knowledge guides the formulation process towards more effective fault isolations. A change in a mental model affects either its structure or behavior. The structure changes when a network branch is pruned from the model and when the model constituents are clustered and decoupled. The behavior changes when a different relevant facet of behavior is selected to be the focus.

Experiential knowledge guiding model refinement consists of:

1. Pattern-recognition rules: Based on the symptom descriptions, the pattern-recognition rules decide which facet of behavior is relevant to the problem at hand. Since these rules also speculate the likely causes of the problem, they indirectly influence the clustering and decoupling of the mental model constituents. Therefore, these rules change the structure and behavior of a mental model.

2. Topological clustering rules: Based on personal experience and preference, the clustering rules decide which adjacent components to agglomerate into clusters. Therefore, these rules change the structure of a mental model.
3. Topological pruning rules: Based on the locality of suspicious components and clusters, the pruning rules decide which branches of the model structure can be pruned without affecting the model simulation. Therefore, these rules change the structure of a mental model.
4. Recommendation rules: Based on domain-specific requirements, the recommendation rules decide which remedial actions to perform in order to correct the individual misbehaviors in the model. Therefore, these rules change the overall behavior of a mental model.

5.10 The DM^2 Algorithm

At the Aspen Security Forum in July 2012, former deputy director of the CIA Counterterrorism Center Henry "Hank" Crumpton spoke of the war on terror: "It's a different type of war. Dealing with terror is going to be more like managing disease" [47].

The generic Dynamic Mental Models (DM^2) can be applied equally well in disease diagnosis as well as counterterrorism. The DM^2 algorithm consists of six major steps (see Fig. 5.1):

1. **Misbehavior pattern recognition**

Based on a body of pattern-recognition rules, DM^2 identifies the misbehavior type for the given symptoms, and speculates the likely sources of the problem. These rules are among the experiential knowledge obtained from a human expert, or they are inferred from statistical data that correlate known symptoms with confirmed causes. In either case, DM^2 collects a list of suspicious components and a list of suspicious clusters for further investigation.

2. **Mental model formulation**

DM^2 constructs a mental model by studying a detailed analytical model of the physical system that it is going to diagnose. Firstly, based on the locality of the suspicious components and clusters, DM^2 prunes the excessive branches from the model structure. This is accomplished with the help of topological pruning rules. Secondly, based on the past experience and personal preference obtained from a human expert, DM^2 clusters the model components in order to increase the mental model abstraction to the level that is more manageable. Thirdly, based on the previously identified misbehavior type, DM^2 chooses to consider only a "relevant facet of behavior" from the complete description of a component.

3. **Mental model refinement**

In the course of diagnosis, DM^2 revises its mental model when necessary. Imagine you are picking up a small box, if you think in your mind that the box is light but it turns out to be heavy, you will be startled for a split second. Your mental model sets up certain expectations that may or may not be met.

The need to modify a mental model arises when the model is not at the right level of abstraction due to new test results and refined speculations. A correct level

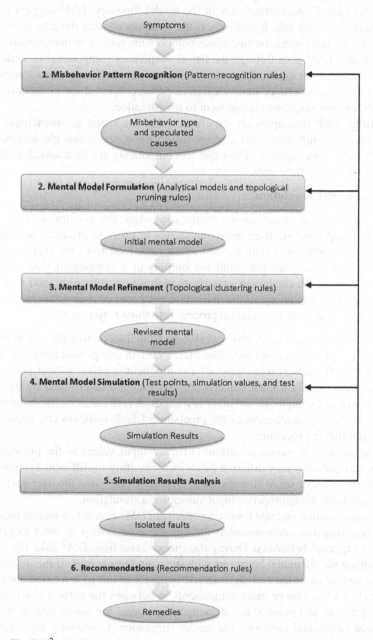

Fig. 5.1 The DM² algorithm flowchart

of model abstraction is essential for effective and efficient diagnosis. To achieve this goal, DM^2 revises its mental model by means of component clustering and decoupling:

(a) Firstly, based on a body of topological clustering rules, DM^2 agglomerates physically adjacent components to form clusters. These rules represent the human expert's perceptive view of the model topology. DM^2 triggers a topological clustering rule if and only if the components that the rule is trying to cluster are not among the suspicious components subject to investigation.

(b) Secondly, DM^2 goes through the list of suspicious components to make sure that they are not embedded inside clusters. For every suspicious component that is hidden in one or more clusters, DM^2 decouples these clusters in order to expose the suspicious component to investigation.

(c) Thirdly, DM^2 decouples all the clusters that are listed as suspicious. In so doing, the components and clusters inside these suspicious clusters become exposed to investigation. The exposed components are then added to the list of suspicious components.

4. **Mental model simulation**

Simulating a mental model means propagating data and constraints from component to component via their interconnections in order to produce the expected behaviors for comparison with the test results obtained from the real system. In other words, we compare the simulated outputs of a component with its actual measurement. For the clusters in a mental model, we consider only their outputs as the net effects of combining all the individual behaviors of their constituents. DM^2 divides the mental model simulation process into three major steps:

(a) Test points restriction: A physical system sometimes limits the extent of testing one can perform and therefore makes certain component behaviors unobservable. The observable output of a component is called a "test point." Test points are usually known beforehand, and they are determined by the availability of test equipment and other resources. A restriction on the testability of physical components reduces the precision of fault isolation and complicates the simulation procedure.

(b) Simulation input values selection: Different input values to the physical system can generate very different behaviors, resulting in different symptoms or none at all under some circumstance. Normally DM^2 asks the user at run time to determine the appropriate input values for a simulation.

(c) Model execution and test results comparison: DM^2 executes a mental model by propagating data and constraints through the components in order to generate their expected behaviors. During the model execution, DM^2 asks the user to compare the simulated behaviors with the actual test results. If the discrepancy between an expected and an observed behavior is out of tolerance, then DM^2 concludes that one or more components in-between the current test point and the previous test point(s) are at fault; it then takes the actual data as the simulated values and continues the model simulation. Continuing the simulation without redoing previous computations and retesting the real system generally saves time and resources when attempting to isolate multiple faults.

5. **Simulation results analysis**

Based on the simulation results from the previous step, DM² decides how to proceed in the diagnosis:

(a) If there is no discrepancy at all between the expected behaviors and the actual test results, DM² gives the user a choice to return to step 4 (Mental Model Simulation) and try different simulation input values. If the repeated attempt fails, DM² suspects incorrect misbehavior-pattern recognition due to misleading symptom descriptions, and it therefore returns to step 1 (Misbehavior Pattern Recognition).

(b) If there is a discrepancy between the expected behaviors and the actual test results, but the fault cannot be isolated, DM² suspects incorrect analytical models, excessive pruning, or inaccurate "relevant facet of behavior", and it returns to step 2 (Mental Model Formulation).

(c) If a cluster is found to be at fault, DM² returns to step 3 (Mental Model Refinement) so that refining the level of abstraction can pinpoint the problems inside the cluster.

(d) Otherwise, DM² proceeds to step 6 (Recommendations).

6. **Recommendations**

Based on a body of recommendation rules obtained from a domain expert, DM² suggests remedial actions for the isolated faults. In many cases, the faulty components are either replaced or adjusted.

5.11 AI Applications in Counterterrorism

After the 9/11 terrorist attacks, more artificial intelligence research and development projects have focused on facilitating knowledge acquisition, assisting in the formation of terrorism-related knowledge bases, and supporting the processes of analysis and decision making in counterterrorism [48].

The University of Arizona's AI Lab developed web-based counterterrorism knowledge portals to "to support the analysis of terrorism research, dynamically model the behavior of terrorists and their social networks, and provide an intelligent, reliable, and interactive communication channel with the terrorized (victims and citizens) groups" [49].

North Atlantic Treaty Organization (NATO) commissioned research on detecting terrorist preparations. FUzzy Signal Expert system for the Detection Of Terrorism preparations (FUSEDOT) applies "artificial intelligence expert system technology to the fuzzy signals presented by certain anomalous data, such as interpersonal relationships, financial relationships, travel patterns, purchasing patterns, patterns of Internet usage, and personal background" [50].

Five U.K. universities launched DScent—a joint project that "combines research theories in the disciplines of computational inference, forensic psychology and expert decision-making in the area of counterterrorism" and it includes the use of neural networks to identify deceptive behavior of terrorists with an average of 60 % success rate [51].

In November 2008, The U.K. government-funded Cyber Security Knowledge Transfer Network (KTN) started examining the potential use of AI in counterterrorism surveillance and data mining. Nigel Jones, director of the KTN, said, "In today's age of distributed networks and with moves towards cloud computing, there is a lot more information out there that might be useful in terms of evidence. There is a problem in handling that mass of data, in storing it, routing it, tracing it and in finding patterns. The [KTN AI and forensics] special interest group will look at the role that AI could have in gathering and analyzing that information, which could be used in investigations or in intelligence gathering to trigger alerts" [52].

In August 2012, Microsoft and NYDP jointly announced bringing the real-time networked Domain Awareness System (DAS) technology to law enforcement agencies around the world. According to retired U.S. Army Lt. Gen. Mike McDuffie, DAS "aggregates and analyzes public safety data in real time and combines artificial intelligence analytics with video from around a jurisdiction to identify potential threats and protect critical infrastructure" [53].

Apart from data mining and public safety, AI will be deployed in the new generation of Unmanned Aerial Vehicles (UAVs) or Unmanned Aerial Systems (UAS) that have become the counterterrorism weapon of choice. In 2009, the U.S. Air Force published "Unmanned Aircraft Systems Flight Plan 2009–2047" which assumes that "the next generation of drones will have artificial intelligence giving them a high degree of operational autonomy including—if legal and ethical questions can be resolved—the ability to shoot to kill" [54].

5.12 Massively Multi-Participant Intelligence Amplification

Hosted by anti-crime activist John Walsh, *America's Most Wanted* premiered in February 1988 on Fox Television Network and moved to Lifetime in December 2011 after 23 years. The TV show profiles and assists law enforcement in the apprehension of fugitives wanted for various crimes. By January 2013, more than 1,200 fugitives have been captured, and over 60 missing children and persons have been found [55]. The success of *America's Most Wanted* is dependent on volunteer citizen detectives among millions of television viewers.

In the computing world, SETI@home (Search for Extraterrestrial Intelligence) was released by UC Berkeley to the public in May 1999 to support the analysis of radio signals from space in an attempt to detect intelligent life outside Earth [56]. The program installs itself as a screensaver and it processes observational data when the home or work computers are idle. By January 2013, SETI@home has a total of 1.3 million users and 3.3 million hosts in 233 countries [57].

In October 2007, Stanford University's Folding@home project received a Guinness World Record for topping 1 petaflop (a thousand trillion floating point operations per second) running on computers as well as Sony's PlayStation 3 video game consoles [58]. Folding@home helps scientists study protein folding and its relationship to Alzheimer's, Huntington's, and cancerous diseases [59]. In September 2011, players of the Foldit video game took less than 10 days to decipher the AIDS-causing Mason-Pfizer monkey virus that had stumped scientists for 15 years [60].

Volunteer citizen detectives and citizen scientists have proved to be immensely successful in data mining and problem solving. We are witnessing the beginning of Massively Multi-Participant Intelligence Amplification (MMPIA) in which a massive network of collective human intelligence, often assisted by distributed computing, is extending the information processing capabilities of the human mind.

In *Facebook Nation: Total Information Awareness*, I proposed a distributed software program called "STS@home" designed for Neighborhood Watch [61]. The artificial intelligence system would analyze live video streams from webcams connected to the homeowners' computers in the Search for Trespassers and Suspects (STS) in their neighborhood. A suspicious activity would trigger an alert, and a neighborhood-watch leader could forward the information to police after reviewing the video clip. A facial recognition feature in the software system could assist police in locating missing children, apprehending fugitives, and solving crimes.

There are already tons of visual information from live traffic cams, ATM security cameras, Neighborhood Watch webcams, and public CCTV systems. Technology exists today that uses video analytics to distill millions of hours of raw video footage into structured, searchable data [62]. Leveraging MMPIA, law enforcement agencies and the intelligence community could conceivably conduct more effective data mining and counterterrorism efforts with the help of citizen detectives and citizen scientists.

References

1. **McCarthy, John.** What is Artificial Intelligence. [Online] Stanford University, November 12, 2007. http://www-formal.stanford.edu/jmc/whatisai/node1.html.
2. **IMDb.** A.I. Artificial Intelligence. [Online] IMDb, June 29, 2001. http://www.imdb.com/title/tt0212720/.
3. **IMDb.** I, Robot. [Online] IMDb, July 16, 2004. http://www.imdb.com/title/tt0343818/.
4. **Paul, Ian.** IBM Watson Wins Jeopardy, Humans Rally Back. [Online] PCWorld, February 17, 2011. http://www.pcworld.com/article/219900/IBM_Watson_Wins_Jeopardy_Humans_Rally_Back.html.
5. **Markoff, John.** Computer Wins on 'Jeopardy!': Trivial, It's Not. [Online] The New York Times, February 16, 2011. http://www.nytimes.com/2011/02/17/science/17jeopardy-watson.html?pagewanted=all.
6. **IBM.** Deep Blue. [Online] IBM. [Cited: November 5, 2012.] http://researchweb.watson.ibm.com/deepblue/.

7. **Goldman, David.** Top U.S. supercomputer guns for fastest in world. [Online] CNNMoney, October 29, 2012. http://money.cnn.com/2012/10/29/technology/innovation/titan-supercomputer/index.html.
8. **Apple.** Learn more about Siri. [Online] Apple. [Cited: November 5, 2012.] http://www.apple.com/ios/siri/siri-faq/.
9. **SRI International.** Cognitive Assistant that Learns and Organizes. [Online] SRI International. [Cited: November 5, 2012.] http://www.ai.sri.com/project/CALO.
10. **Markoff, John.** A Software Secretary That Takes Charge. [Online] The New York Times, December 13, 2008. http://www.nytimes.com/2008/12/14/business/14stream.html.
11. **Hay, Timothy.** Apple Moves Deeper Into Voice-Activated Search With Siri Buy. [Online] The Wall Street Journal, April 28, 2010. http://blogs.wsj.com/venturecapital/2010/04/28/apple-moves-deeper-into-voice-activated-search-with-siri-buy/.
12. **Minsky, Marvin.** THE AGE of INTELLIGENT MACHINES | Thoughts About Artificial Intelligence. [Online] Kurzweil Accelerating Intelligence, February 21, 2001. http://www.kurzweilai.net/marvin-minsky.
13. **CNN.** AI set to exceed human brain power. [Online] CNN, August 9, 2006. http://www.cnn.com/2006/TECH/science/07/24/ai.bostrom/.
14. **3VR Inc.** Use Video Analytics and Data Decision Making to Grow Your Business. [Online] Digital Signage Today. [Cited: May 28, 2012.] http://www.digitalsignagetoday.com/whitepapers/4891/Use-Video-Analytics-and-Data-Decision-Making-to-Grow-Your-Business.
15. **Rayner, Gordon.** New intelligent CCTV cameras can see and hear. [Online] The Telegraph, June 23, 2008. http://www.telegraph.co.uk/news/uknews/2180628/New-intelligent-CCTV-cameras-can-see-and-hear.html.
16. **Aviv, Juval.** Can AI Fight Terrorism? [Online] Forbes, June 22, 2009. http://www.forbes.com/2009/06/18/ai-terrorism-interfor-opinions-contributors-artificial-intelligence-09-juval-aviv.html.
17. **New York City Police Department.** Public Security Privacy Guidelines. [Online] New York City Police Department, April 2, 2009. http://www.nyc.gov/html/nypd/downloads/pdf/crime_prevention/public_security_privacy_guidelines.pdf.
18. **Thistle, Michele Bedford.** Microsoft, NYPD team up on global counterterrorism solution. [Online] Microsoft, August 9, 2012. http://www.microsoft.com/government/ww/public-services/blog/Pages/post.aspx?postID=158&aID=4.
19. **Bamford, James.** The NSA Is Building the Country's Biggest Spy Center (Watch What You Say). [Online] Wired, March 15, 2012. http://www.wired.com/threatlevel/2012/03/ff_nsadatacenter/all/.
20. **Bamford, James.** War of Secrets; Eyes in the Sky, Ears to the Wall, and Still Wanting. [Online] The New York Times, September 8, 2002. http://www.nytimes.com/2002/09/08/weekinreview/war-of-secrets-eyes-in-the-sky-ears-to-the-wall-and-still-wanting.html?pagewanted=all.
21. **Bamford, James.** War of Secrets; Eyes in the Sky, Ears to the Wall, and Still Wanting. [Online] The New York Times, September 8, 2002. http://www.nytimes.com/2002/09/08/weekinreview/war-of-secrets-eyes-in-the-sky-ears-to-the-wall-and-still-wanting.html?pagewanted=all.
22. **Kalil, Tom.** Big Data is a Big Deal. [Online] The White House, March 29, 2012. http://www.whitehouse.gov/blog/2012/03/29/big-data-big-deal.
23. **Chakrabarti, Soumen, et al.** Data Mining Curriculum: A Proposal (Version 1.0). [Online] ACM SIGKDD, April 30, 2006. http://www.sigkdd.org/curriculum/CURMay06.pdf.
24. **Block, Robert, Fields, Gary and Wrighton, Jo.** U.S. 'Terror' List Still Lacking. [Online] The Wall Street Journal, January 2, 2004. http://online.wsj.com/article/0,,SB107300534268021800,00.html.
25. **Gorman, Siobhan.** How Team of Geeks Cracked Spy Trade. [Online] The Wall Street Journal, September 4, 2009. http://online.wsj.com/article/SB125200842406984303.html.
26. **Flynn, Michael T.** Advanced Analytical Capability Joint Urgent Operational Need Statement. [Online] Department of Defense US Forces Afghanistan, July 2, 2010. http://www.politico.com/static/PPM223_110629_flynn.html.

27. **Carik, Kenneth.** The Nature of Explanation. [Online] Cambridge University Press, UK, 1943. http://www.cambridge.org/us/knowledge/isbn/item1121731/.
28. **Minsky, Marvin.** A Framework for Representing Knowledge. [Online] MIT AI Laboratory Memo 306, June 1974. http://web.media.mit.edu/~minsky/papers/Frames/frames.html.
29. **DeRosa, Mary.** Data Mining and Data Analysis for Counterterrorism. [Online] Center for Strategic and International Studies, March 2004. http://csis.org/files/media/csis/pubs/040301_data_mining_report.pdf.
30. **McCarthy, John.** Recursive Functions of Symbolic Expressions and Their Computation by Machine, Part I. [Online] Massachusetts Institute of Technology, April 1960. http://www-formal.stanford.edu/jmc/recursive/recursive.html.
31. **Simran, Max & Charence.** Prolog. [Online] Imperial College London, 2006. http://www.doc.ic.ac.uk/~cclw05/topics1/prolog.html.
32. **Brachman, Ronald J. and Schmolze, James G.** An Overview of the KL-ONE Knowledge Representation System. [Online] Cognitive Science, 1985. http://eolo.cps.unizar.es/docenci a/MasterUPV/Articulos/An%20Overview%20of%20the%20KL-ONE%20Knowledge%20 Representation%20System-Brachman1985.PDF.
33. **Minsky, Marvin.** Logical vs. Analogical or Symbolic vs. Connectionist or Neat vs. Scruffy. [Online] MIT Press, 1990. http://web.media.mit.edu/~minsky/papers/SymbolicVs.Connectio nist.html.
34. **Liu, Ling and Ozsu, M. Tamer.** Encyclopedia of Database Systems. [Online] Springer, 2009. http://www.springer.com/computer/database+management+%26+information+retrie val/book/978-0-387-35544-3.
35. **Wu, Xindong.** Knowledge Acquisition from Databases. [Online] Ablex Publishing Corporation, 1995. http://www.cs.uvm.edu/~xwu/Publication/Book-95.html.
36. **Buchanan, Bruce G. and Shortliffe, Edward H.** Rule-Based Expert Systems: The MYCIN Experiments of the Stanford Heuristic Programming Project. [Online] Addison Wesley, 1984. http://www.amia.org/staff/eshortliffe/Buchanan-Shortliffe-1984/MYCIN%20Book.htm.
37. **Lee, Newton S., Phadke, Madhav S. and Keny, Rajiv.** An expert system for experimental design in off-line quality control. [Online] Wiley, November 1989. http://onlinelibrary.wiley.com/doi/10.1111/j.1468-0394.1989.tb00148.x/abstract.
38. **Chessbase.** Dark horse ZackS wins Freestyle Chess Tournament. [Online] Chessbase News, June 19, 2005. http://www.chessbase.com/newsdetail.asp?newsid=2461.
39. **Lee, Newton S.** DM2: an algorithm for diagnostic reasoning that combines analytical models and experiential knowledge. [Online] International Journal of Man-Machine Studies, June 1988. http://www.sciencedirect.com/science/article/pii/S002073738880066X.
40. **De Kleer, Johan.** How Circuits Work. [Online] Artificial Intelligence, December 1984. http://www2.parc.com/spl/members/dekleer/Publications/How%20Circuits%20Work.pdf.
41. **Genesereth, Michael R.** The use of design descriptions in automated diagnosis. [Online] Artificial Intelligence, December 1984. http://www.sciencedirect.com/science/article/pii/0004370284900432.
42. **Davis, Randall.** Diagnostic reasoning based on structure and behavior. [Online] Artificial Intelligence, December 1984. http://www.sciencedirect.com/science/article/pii/0004370284900420.
43. **Berwaner, Mary.** The Problem of Diagnostic Aiding. [Online] The Defense Technical Information Center, October 30, 1989. http://www.dtic.mil/cgi-bin/GetTRDoc?AD=ADA239200.
44. **Trafton, J. Gregory.** Dynamic mental models in weather forecasting. [Online] Defense Technical Information Center, 2004. http://www.dtic.mil/cgi-bin/GetTRDoc?AD=ADA480241.
45. **Gentner, Dedre and Stevens, Albert L.** Mental Models. [Online] Lawrence Erlbaum Associates, May 1, 1983. http://books.google.com/books/about/Mental_Models.html? id=QFI0SvbieOcC.
46. **Williams, Michael D. and Hollan, James D., Stevens, Albert L.** Human Reasoning About a Simple Physical System. [Online] Lawrence Erlbaum Associates, May 1, 1983. http://books.google.com/books?id=QFI0SvbieOcC&pg=PA131.
47. **Dougherty, Jill.** Experts: No easy cure for the disease of terror. [Online] CNN, July 27, 2012. http://security.blogs.cnn.com/2012/07/27/experts-no-easy-cure-for-the-disease-of- terror/.

48. **Markman, A.B., et al.** Analogical Reasoning Techniques In Intelligent Counterterrorism Systems. [Online] International Journal on Information Theories and Applications, 2003. http://www.foibg.com/ijita/vol10/ijita10-2-p04.pdf.

49. **Reid, Edna, et al.** Terrorism Knowledge Discovery Project: A Knowledge Discovery Approach to Addressing the Threats of Terrorism. [Online] The University of Arizona. [Cited: November 24, 2012.] http://ai-vm-s08-rs1-1.ailab.eller.arizona.edu/people/edna/AILab_terrorism%20Knowledge%20Discovery%20ISI%20_apr04.pdf.

50. **Koltko-Rivera, Mark E.** Detection of Terrorist Preparations by an Artificial Intelligence Expert System Employing Fuzzy Signal Detection Theory. [Online] Defense Technical Information Center, October 2004. http://www.dtic.mil/cgi-bin/GetTRDoc?AD=ADA460204.

51. **Dixon, S.J., et al.** Neural Network for Counter-Terrorism. [Online] Leeds Metropolitan University. [Cited: November 24, 2012.] http://www.leedsmet.ac.uk/aet/computing/aNNforCTShortPaper.pdf.

52. **Heath, Nick.** Police enlist AI to help tackle crime. [Online] ZDNet, November 6, 2008. http://www.zdnet.com/police-enlist-ai-to-help-tackle-crime-3039541531/.

53. **McDuffie, Mike.** Microsoft and NYPD Announce Partnership Providing Real-Time Counterterrorism Solution Globally. [Online] Microsoft, August 8, 2012. http://www.microsoft.com/government/en-us/state/brightside/Pages/details.aspx?Microsoft-and-NYPD-Announce-Partnership-Providing-Real-Time-Counterterrorism-Solution-Globally&blogid=697.

54. **The Economist.** Flight of the drones: Why the future of air power belongs to unmanned systems. [Online] The Economist, October 8, 2011. http://www.economist.com/node/21531433.

55. **America's Most Wanted.** America's Most Wanted. [Online] Lifetime. [Cited: January 17, 2013.] http://www.amw.com/.

56. **SETI@home.** SETI@home. [Online] University of California. [Cited: January 17, 2013.] http://setiathome.berkeley.edu/index.php.

57. **BOINC STATS.** Project stats info. *BOINC STATS.* [Online] January 15, 2013. http://boincstats.com/en/stats/projectStatsInfo.

58. **Terdiman, Daniel.** Sony's Folding@home project gets Guinness record. [Online] CNet, October 31, 2007. http://news.cnet.com/8301-13772_3-9808500-52.html.

59. **Stanford University.** Folding@home distributed computing. [Online] Stanford University. [Cited: January 17, 2013.] http://folding.stanford.edu/English/HomePage.

60. **Boyle, Alan.** Gamers solve molecular puzzle that baffled scientists. [Online] NBC News, September 18, 2011. http://cosmiclog.nbcnews.com/_news/2011/09/18/7802623-gamers-solve-molecular-puzzle-that-baffled-scientists.

61. **Lee, Newton.** Facebook Nation: Total Information Awareness. [Online] Springer, September 15, 2012. http://www.amazon.com/Facebook-Nation-Total-Information-Awareness/dp/1461453070.

62. **3VR.** Use Video Analytics and Data Decision Making to Grow Your Business. [Online] [Cited: May 28, 2012.] http://www.digitalsignagetoday.com/whitepapers/4891/Use-Video-Analytics-and-Data-Decision-Making-to-Grow-Your-Business.

Part III
Counterterrorism Technologies:
Social Media and Cybersecurity

Part III

Chapter 6
Social Media and Two-Way Street of Total Information Awareness

> *The fantasy worlds that Disney creates have a surprising amount in common with the ideal universe envisaged by the intelligence community, in which environments are carefully controlled and people are closely observed, and no one seems to mind.*
> —Lawrence Wright, *The New Yorker* (January 21, 2008).

> *Our job as citizens is to ask questions.*
> —Thomas Blanton, National Security Archive.
> George Washington University (December 16, 2010).

> *The two-way street of Total Information Awareness is the road that leads to a more transparent and complete picture of ourselves, our governments, and our world.*
> —Newton Lee.

6.1 It's a Small World, with CCTVs

In January 2008, Pulitzer Prize-winner Lawrence Wright wrote in *The New Yorker* an in-depth article about the U.S. intelligence community focusing on the Office of the Director of National Intelligence (ODNI) and the necessity for interagency communications—something that Total Information Awareness (TIA) was meant to facilitate. Wright observed that "the fantasy worlds that Disney creates have a surprising amount in common with the ideal universe envisaged by the intelligence community, in which environments are carefully controlled and people are closely observed, and no one seems to mind" [1].

In addition to the bag checks at the entrances to Disney theme parks, plain clothes security officers and closed-circuit television (CCTV) hidden cameras have kept the parks safe without intruding on the privacy of the guests. Other than a few rare incidents, Disneyland is "the happiest place on earth" [2].

N. Lee, *Counterterrorism and Cybersecurity*, DOI: 10.1007/978-1-4614-7205-6_6, © Springer Science+Business Media New York 2013

In the year 2012, from ATMs to parking lots to shopping malls, there are approximately 30 million cameras in the world capturing 250 billion hours of raw footage annually [3]. In the United Kingdom, CCTV is so prevalent that some residents can expect to be captured by a camera at least 300 times a day [4]. With more than 1.85 million cameras operating in the U.K. [5], the security-camera cordon surrounding London has earned the nickname of "Ring of Steel" [6]. The U.K. first introduced the security measures in London's financial district in mid-1990s during an Irish Republican Army (IRA) bombing campaign. After the 9/11 terrorist attacks, the "Ring of Steel" was widened to include more businesses [7].

Since the 1970s, the proliferation of CCTV cameras in public places has led to some unease about the erosion of civil liberties and individual human rights, along with warnings of an Orwellian "Big Brother" culture. Nevertheless, nowadays we all have accepted the presence of CCTV in public places.

In the U.S., New York, Los Angeles, San Francisco, and Chicago are among the major cities that have implemented citywide CCTV monitoring systems. Disney theme parks, Six Flags, and other public attractions also use video surveillance systems that can see in the dark.

Tourists not only love to visit Disneyland but also flock to Las Vegas casinos and resorts, another fantasy world, where security cameras are in ample use. In March 2012, Mirage Resort in Las Vegas became the 50th casino to install facial recognition software as part of the surveillance suite of Visual Casino loss-reduction systems [8] .

6.2 Facebook Nation: Total Information Awareness

President Barack Obama, in his 2011 State of the Union Address, called America "the nation of Edison and the Wright brothers" and "of Google and Facebook" [9]. Enormous amounts of information are being gathered on everyone living in the Facebook nation. For the 2012 presidential election, Obama's data-mining team created a massive database of voter information, consumer data, and social media contacts [10]. The analysis of big data enabled Obama's campaign to run computer simulations, fundraise a staggering $1 billion dollars, reach the swing-state voters more effectively, and ultimately win the reelection for President Obama.

In a pep talk at Wakefield High School in September 2009, Obama told the students, "Be careful what you post on Facebook. Whatever you do, it will be pulled up later in your life" [11]. In August 2012, Prof. Amitai Etzioni of George Washington University opined that "Facebook merely adds to the major inroads made by the CCTV cameras that are ubiquitous in many cities around the globe, along with surveillance satellites, tracking devices, spy malware and, most recently, drones used not for killing terrorists but for scrutinizing spaces heretofore considered private, like our backyards. Corporations keep detailed dossiers on what we purchase. No wonder privacy advocates argue that we live in a surveillance society and privacy 'ended with Facebook'" [12].

A year after TIA was officially shut down in 2003, Facebook was born in 2004 with about 650 users during its first week of debut. In August 2008, Facebook had grown to 100 million users. By July 2010, Facebook reached 500 million. And in October 2012, Facebook topped 1 billion monthly active users [13]. As a nation, Facebook would be the third largest country in the world with over 1 billion citizens, after China and India.

In an interview with *Ad Age*'s Ann-Christine Diaz, Facebook's head of consumer marketing Rebecca Van Dyck linked Facebook with the innate human desire to connect. Dyck said, "We make the tools and services that allow people to feel human, get together, open up. Even if it's a small gesture, or a grand notion— we wanted to express that huge range of connectivity and how we interact with each other" [14].

On October 4, 2012, Facebook released a new 91-second video *The Things That Connect Us* depicting chairs, doorbells, airplanes, bridges, dance floors, basketball, a great nation, and the universe [15].

> Chairs. Chairs are made so that people can sit down and take a break. Anyone can sit on a chair, and if the chair is large enough, they can sit down together. And tell jokes. Or make up stories. Or just listen. Chairs are for people. And that is why **chairs are like Facebook**.
>   Doorbells. Airplanes. Bridges. These are things people use to get together so they can open up and connect about ideas, and music, and other things people share.
>   Dance floors. Basketball. A great nation. A great nation is something people build, so that they can have a place where they belong.
>   The universe is vast and dark and makes us wonder if we are alone. So maybe the reason we make all of these things is to remind ourselves that we are not.

Directed by acclaimed Mexican filmmaker Alejandro González Iñárritu, the cleverly crafted video has been described by some critics as "puzzling" and "disingenuous" [16]. Nonetheless, it is not difficult to see that the video alludes to the rise of Facebook nation with over 1 billion cybercitizens. It is truly a global phenomenon since the majority of Facebook users (81 %) live outside the U.S. and Canada [13].

"Chairs are like Facebook"—Chairs are the most basic, ubiquitous, and indispensable furniture in most parts of the world. Facebook is one of the most prevalent social networks today. However, sitting in stationary chairs puts stress on spinal disks and increases the chance of lower-back injury, resulting in $11 billion a year in workers' compensation claims [17]. Unlike stationary chairs, Facebook must be quick to adapt to changes.

"We are not [alone]"—Facebook users tell jokes, make up stories, or just listen to other Facebook friends. In my 2012 book *Facebook Nation: Total Information Awareness*, I portray the social media ecosystem as a world of increasing total information awareness, which is essentially a civilian version of Poindexter's TIA program [18]. On Facebook, people volunteer their personal information such as their gender, birthday, education, workplace, city of residence, interests, hobbies, photos, friends, families, schoolmates, coworkers, past histories, relationship status, likes, dislikes, and even current location. WikiLeaks founder Julian Assange told RT's Laura Emmett in a May 2011 interview [19]:

Facebook in particular is the most appalling spying machine that has ever been invented. Here we have the world's most comprehensive database about people: their relationships, their names, their addresses, their locations and their communications with each other, their relatives—all sitting within the United States, all accessible to US intelligence. Facebook, Google, Yahoo!—all these major US organizations have built-in interfaces for US intelligence. It's not a matter of serving a subpoena. They have an interface that they have developed for U.S. intelligence to use.

Although Assange's bold accusation of Facebook and social media is subject to debate, it is open knowledge that law enforcement authorities in New York, Atlanta, San Diego, and Chicago have been using Facebook to gather evidence against gang members and criminals. The success of social media sleuthing has prompted the New York Police Department (NYPD) to double the size of its online investigators in October 2012.

"By capitalizing on the irresistible urge of these suspects to brag about their murderous exploits on Facebook, detectives used social media to draw a virtual map of their criminal activity over the last three years," said NYPD commissioner Raymond Kelly [20].

Donna Lieberman, executive director of the New York Civil Liberties Union, concurred with Kelly. Lieberman said, "NYPD has the right, indeed the obligation, to pursue effective avenues for investigating criminal gang activity, and that includes using Facebook and other social media. But such methods must be closely monitored so they don't become a vehicle for entrapment or unauthorized surveillance" [20].

The truth of the matter is that there is hardly any private information on the Internet. The U.S. Library of Congress has been archiving Web content since 2000. Twitter and the federal library announced in April 2010 that every public tweet posted since 2006 would be archived digitally [21]. By January 2013, the Library of Congress has compiled a total of more than 170 billion Twitter messages and it is now processing about 500 million new tweets per day [22]. The federal library plans to make all the tweets available to researchers and the general public in the near future.

6.3 Surveillance Satellites, Tracking Devices, Spy Malware, and Drones

A year before Prof. Amitai Etzioni's wrote the *CNN* article in 2012 about "surveillance satellites, tracking devices, spy malware, and drones used not for killing terrorists but for scrutinizing spaces heretofore considered private, like our backyards" [12], the Defense Advanced Research Projects Agency (DARPA) and AeroVironment had developed the Nano Hummingbird—a miniature drone in camouflage that looks like a hummingbird capable of maneuvering in urban areas [23]. If we really take a look around, we can find ourselves surrounded by:

1. **Surveillance satellites for Google Earth and Maps**

Satellite images provide the necessary database for Google Earth, Google Maps, and other useful applications. At the "Next Dimension" Google Maps press event in June 2012, Google announced that it has over 1 billion monthly users for all of Google Map Services [24]. In July 2012, Google published new high-resolution aerial and satellite imagery for 25 cities and 72 countries/regions in both Google Earth and Maps.

Geo Data Specialist Bernd Steinert wrote in a Google blog, "In our continuing effort to build the most comprehensive and accurate view of the world, the Google Earth and Maps Imagery team just published another extensive catalog of new imagery. This week we have exciting new updates to both our high resolution aerial and satellite imagery and our 45° imagery" [25].

Google Earth and Maps enable us to see not only the amazing world but also our neighbors' private backyards.

2. **Tracking devices such as iPhone, iPad and Carrier IQ**

In April 2011, O'Reilly Radar reported that iPhones and 3G iPads are regularly recording the position of the device into a hidden file called "consolidated.db" [26]. The secret database file has been storing the locations (latitude-longitude coordinates) and time stamps, effectively tracking the history of movement of the iPhone and 3G iPad users for a year since iOS 4 was released in 2010.

Not to be outdone by the iPhone location tracking software, the Carrier IQ software has been found on about 150 million cell phones including the iPhones, Android, BlackBerry, and Nokia phones [27]. On November 28, 2011, security researcher Trevor Eckhart posted a video on YouTube detailing hidden software installed on smartphones that secretly logs keypresses, SMS messages, and browser URLs [28].

Our trusted smartphones have unknowingly become the tracking devices for big businesses.

3. **Spy malware using cookies, tracking code, and mobile apps**

In February 2012, Jonathan Mayer, a graduate student at Stanford University, demonstrated that four advertising companies, Google's DoubleClick, Vibrant Media, Media Innovation Group, and PointRoll, have been deliberately circumventing Apple Safari's privacy feature by installing temporary cookies on the user devices in order to track users' behavior [29]. Safari is the primary web browser on the iPhone, iPad, and Macintosh computers. The Stanford findings contradicted Google's own instructions to Safari users on how to avoid tracking.

According to a *Wall Street Journal* Research conducted by Ashkan Soltani, Google placed the tracking code within ads displayed on 29 of the top 100 most-visited U.S. websites [30]. Among them are household names YouTube, AOL,

People Magazine, New York Times, WebMD, Merriam-Webster Dictionary, Fandango.com, Match.com, TMZ, and Yellow Pages.

Also in February 2012, Twitter acknowledged that when a user taps the "Find friends" feature on its smartphone app, the company downloads the user's entire address book, including email addresses and phone numbers, and keeps the data on its servers for 18 months [31].

Off-the-shelf anti-virus software can protect us from malware but not the sophisticated spyware from trusted companies.

4. Drones operated by police, civilians, and Google

Police departments in Seattle, Miami, Little Rock, and other cities have been using unmanned drones for surveillance and law enforcement purposes. According to *U.S. News and World Report,* drones are used to "gain an aerial perspective consistent with the open view doctrine," which allows officers to monitor areas that are in "plain view" [32].

Congressman Hank Johnson of Georgia voiced his concerns, "As the number of drones rises, so, too, will the number of suspects. During the civil rights movement, would activists have left their homes if they knew they were being monitored from cameras 30,000 feet above" [33]?

On the flip side, low-cost drones under $300 each enabled protestors in Occupy Wall Street to monitor the police. In December 2011, activists remotely piloted a Parrot AR.Drone, dubbed "The Occucopter," on their smartphones to provide a live feed of Occupy Wall Street from above [34]. An increasing number of civilians including journalists, wildlife researchers, sports photographers, and real estate agents are using drones for their work [35].

Nonetheless, police and civilian drones pale in comparison to the ubiquitous Google Street View. Launched in May 2007, Google Street View has captured 20 petabytes of data in 39 countries and about 3,000 cities [36]. Google uses cars, trikes, boats, snowmobiles, trolleys, and people outfitted with custom cameras to capture 360° panoramic images around the world.

In May 2010, however, Google made a stunning admission that for over three years, its camera-toting Street View cars have inadvertently collected snippets of private information that people send over unencrypted WiFi networks [37]. In October 2010, Google admitted to "accidentally" collecting and storing entire e-mails, URLs, and passwords from unsecured WiFi networks with its Street View cars in more than 30 countries, including the United States, Canada, Mexico, some of Europe, and parts of Asia [38].

At the "Next Dimension" Google Maps press event in June 2012, Google Street View engineering director Luc Vincent demonstrated a new prototype backpack with camera rig. Vincent explained that "the camera has 15, 5-megapixel lenses and the battery for the rig lasts all day." But he also quipped that "if you hike with a partner, they'll be in every scene" [36].

In October 2012, Google Street View launched its biggest update ever by doubling the number of special collections, adding 250,000 miles of roads around the world, and increasing Street View coverage in Macau, Singapore, Sweden, the U.S., Thailand, Taiwan, Italy, the United Kingdom, Denmark, Norway, and Canada.

Google Street View Program Manager Ulf Spitzer said, "Street View, as you know, is a useful resource when you're planning a route or looking for a destination, but it can also magically transport you to some of the world's picturesque and culturally significant landmarks" [39]. Google Street View is indeed handy for planning a trip to a new restaurant or an unfamiliar neighborhood as well as exploring national parks, university campuses, sports stadiums, and museums around the world. We are living in an information age where abundant data are readily available at our fingertips, whether we are individuals, businesses, or government agencies.

"I fear Google more than I pretty much fear the government," said Jeff Moss, founder of Black Hat and DEF CON, at the 2012 Black Hat Conference. "Google, I'm contractually agreeing to give them all my data" [40].

6.4 Two-Way Street of Total Information Awareness

Private businesses and the ubiquity of social networks are creating the necessary technologies and infrastructures for Total Information Awareness (TIA) [18]. However, unlike the government-proposed TIA which is a rigid one-way mirror, the industry-led TIA is an evolving two-way street. Facebook, Google, YouTube, Wikipedia, and even the controversial WikiLeaks all collect information from everywhere and make it available to everyone, whether they are individuals, businesses, or government agencies.

There are some safety concerns about exposing too much information online, especially high value terrorist targets. According to the Federal Bureau of Investigation (FBI), the individuals who planned the attempted car bombing of Times Square on May 1, 2010 used public web cameras for reconnaissance [41].

As a precaution, Google Maps has digitally modified or blurred its satellite imagery on some landmarks including the roof of the White House, NATO air force hub serving as a retreat for the Operation Iraqi Freedom forces, Mobil Oil facilities in Buffalo New York, and even the Dutch Royal Family's Huis Ten Bosch Palace in Netherlands [42].

Pete Cashmore, founder and CEO of *Mashable,* suggested that the world of 2012 is both reminiscent of George Orwell's 1984 and radically at odds with it. Cashmore wrote in a January 2012 *CNN* article, "The online world is indeed allowing our every move to be tracked, while at the same time providing a counterweight to the emergence of Big Brother... Unlike in Orwell's dystopian world, people today are making a conscious choice to do so. The difference between this reality and Orwell's vision is the issue of control: While his Thought Police tracked you without permission, some consumers are now comfortable with sharing their every move online" [43].

Not only are consumers knowingly sharing their private information with big businesses, government officials are also leaking classified information to the media and private companies for either financial gains or political purposes. In 2012, seven U.S. Navy SEALs were reprimanded for divulging classified combat gear to Electronic Arts, maker of the multiplatform game "Medal of Honor: Warfighter" [44]. A string of leaks from high-ranking government officials resulted in the public airing of details about the U.S. cyber attack on Iran's nuclear centrifuge program [45], increased U.S. drone strikes against militants in Yemen [46] and Pakistan [47], and a double agent disrupting al-Qaeda's custom-fit underwear bombing plot [48].

Even President Barack Obama himself revealed on a Google+ video chat room interview in January 2012 that the U.S. conducted many drone strikes to hunt down al-Qaeda and Taliban in Pakistan [49]. Some former military and intelligence officers accused Obama of disclosing clandestine operations to the public. Ex-Navy SEAL Benjamin Smith voiced his opposition to leaks, "As a citizen, it is my civic duty to tell the president to stop leaking information to the enemy. It will get Americans killed." Another former Navy SEAL Scott Taylor said of the bin Laden raid, "If you disclose how we got there, how we took down the building, what we did, how many people were there, that it's going to hinder future operations, and certainly hurt the success of those future operations"[50].

Then CIA director David Petraeus issued a statement to his employees in January 2012 about the arrest of a former Agency officer by the FBI on charges that he illegally disclosed classified information to reporters. Petraeus wrote, "Unauthorized disclosures of any sort—including information concerning the identities of other Agency officers—betray the public trust, our country, and our colleagues" [51].

On the other hand, Thomas Blanton, Director of National Security Archive at George Washington University, testified in December 2010 before the U.S. House of Representatives on the massive overclassification of the U.S. government's national security information. At the Judiciary Committee hearing on the Espionage Act and the legal and constitutional implications of WikiLeaks, Blanton said that "Actually our job as citizens is to ask questions. Experts believe 50–90 % of our national security secrets could be public with little or no damage to real security" [52]. He cited his findings:

A few years back, when Rep. Christopher Shays (R-CT) asked Secretary of Defense Donald Rumsfeld's deputy for counterintelligence and security how much government information was overclassified, her answer was 50 %. After the 9/11 Commission reviewed the government's most sensitive records about Osama bin Laden and Al-Qaeda, the co-chair of that commission, former Governor of New Jersey Tom Kean, commented that "three-quarters of what I read that was classified shouldn't have been"—a 75 % judgment. President Reagan's National Security Council secretary Rodney McDaniel estimated in 1991 that only 10 % of classification was for "legitimate protection of secrets"—so 90 % unwarranted. Another data point comes from the Interagency Security Classification Appeals Panel, over the past 15 years, has overruled agency secrecy claims in whole or in part in some 65 % of its cases.

WikiLeaks founder Julian Assange said in a 2011 interview, "Our No. 1 enemy is ignorance. And I believe that is the No. 1 enemy for everyone—it's not understanding what actually is going on in the world" [53]. Bill Keller, former executive editor of *The Times*, opines that "the most palpable legacy of the WikiLeaks campaign for transparency is that the U.S. government is more secretive than ever" [54]. U.S. Air Force Senior Airman Christopher R. Atkins wrote in his email to American filmmaker Michael Moore, "The single greatest danger to America and our way of life is ourselves. No foreign power can dictate your oppression. No foreign army can impose martial law upon us. No foreign dictator can remove the precious right that I am exercising at this moment. Militaries do not keep people free! Militaries keep us safe, but it is we citizens who ensure freedom!" [55].

Notwithstanding the potential risks and benefits of information sharing, the two-way street of Total Information Awareness is the road that leads to a more transparent and complete picture of ourselves, our governments, and our world. As Wikipedia's founder Jimmy Wales said, "Imagine a world in which every single person on the planet is given free access to the sum of all human knowledge" [56].

6.5 No Doomsday for the Internet

Governments around the world convened in Dubai for the World Conference on International Telecommunications (WCIT) from December 3 to 14, 2012 to decide on the future of the open Internet [57]. Hosted by United Nations' International Telecommunication Union (ITU), the conference was to update the 1988 International Telecommunication Regulations (ITRs) behind closed doors. However, documents have leaked, showing some government proposals that may threaten the open Internet [58]. Among them were government's "right to know how its traffic is routed" and restrictions on public access to telecommunications when used for "undermining the sovereignty, national security, territorial integrity and public safety of other States, or to divulge information of a sensitive nature."

"Proposals by various governments to treat internet connections like the telephone system are cause for concern regarding privacy and the unfettered, free flow of information," said Emma Llansó, policy counsel at the Center for Democracy and Technology. "[But] there's not going to be some kind of doomsday scenario that there's a treaty that makes the internet go dark. What we're seeing is governments putting forward visions of the internet and having discussions" [59].

The Egyptian government shut down Internet access at midnight January 28, 2011 when activists organized through Facebook and Twitter nationwide protests to call for an end to President Hosni Mubarak's government [60]. Google responded to the Internet blockade by working with Twitter and SayNow to unveil a web-free speak-to-tweet service, allowing anyone to send and receive tweets by calling a phone number [61]. The unprecedented totalitarian action failed to thwart the planned demonstrations, and Mubarak resigned as president.

To disrupt rebel communications, the Syrian government shut down the Internet across the country and cut cellphone services in select areas on November 29, 2012. *The Associated Press* reported that "the revolt in Syria began with peaceful protests but turned into a civil war after the government waged a brutal crackdown on dissent" [62]. As the Arab Spring uprising continues to spread throughout the Middle East region, governments should have learned that it is futile to suppress the public or enemies by shutting down the Internet and social networks. Doing so only ends up backfiring.

President Ronald Reagan had predicted that "technology will make it increasingly difficult for the state to control the information its people receive" [63]. He said in June 1989, "Information is the oxygen of the modern age. It seeps through the walls topped by barbed wire, it wafts across the electrified borders, the Goliath of totalitarianism will be brought down by the David of the microchip" [64].

Terry Kramer, U.S. Ambassador to 2012 WCIT, refused to sign the United Nations (U.N.) treaty on telecommunications and Internet. He said, "The Internet has given the world unimaginable economic and social benefit during these past 24 years, all without U.N. regulation" [65].

Nevertheless, the OpenNet Initiative (ONI) reported in April 2012 that 42 countries filter and censor content out of the 72 studied, while 21 countries have been engaging in substantial or pervasive filtering [66]. Freedom House published its findings that "even a number of democratic states have considered or implemented various restrictions in response to the potential legal, economic, and security challenges raised by new media" [67].

Vinton Cerf, a father of the Internet and Google's chief Internet evangelist, strongly opposes government intervention in the World Wide Web. He said that the Internet allows "all of us to reach a global audience at a click of a mouse," and he warned that "history is rife with examples of governments taking actions to 'protect' their citizens from harm by controlling access to information and inhibiting freedom of expression and other freedoms outlined in The Universal Declaration of Human Rights. We must make sure, collectively, that the Internet avoids a similar fate" [68].

6.6 Web 2.0 for Intelligence Community: Intellipedia, A-Space, Deepnet

In 2004, Calvin Andrus of the Central Intelligence Agency (CIA) wrote a paper entitled "The Wiki and the Blog: Toward a Complex Adaptive Intelligence Community," calling for the U.S. intelligence community to follow the lead of the social media industry. In 2006, Sean Dennehy and Don Burke oversaw the creation of Intellipedia, the U.S. intelligence community's version of Wikipedia on top secret, secret, and unclassified networks.

Burke said, "In addition to analysis, we need people who can create an ecosystem of knowledge that is not specifically about answering tomorrow's questions, but creating a world of information that is connected" [69].

Dennehy added, "There's too much emphasis on the analytical report. It's important to look at how we get to the finished intelligence. Intellipedia does this by making the process more social and creating a dialogue that's transparent" [69].

In 2008, the U.S. intelligence community launched A-Space (Analytic Space), an online collaboration environment modeled after MySpace and Facebook to improve analysts' abilities to share information, form communities, and collaborate [70].

"After an initial slow start in which analysts were skeptical about sharing information, they now see that collaboration reduces the time and effort they spend on analyzing data," said Michael Wertheimer of the Office of the Director of National Intelligence (ODNI) [71].

A 2010 Defense Science Board reported that the National Security Agency (NSA) has been developing tools to find and index data in the deep web (aka deepnet) that "contains government reports, databases, and other sources of information of high value to Department of Defense (DoD) and the intelligence community" [72].

It should not come as a surprise that social networking is becoming a tool of espionage. In 2009, CIA double agent Morten Storm was tasked to ensnare the American-born al-Qaeda cleric Anwar al-Awlaki, one of the most wanted terrorists. Dubbed the "bin Laden of the Internet," al-Awlaki had a blog, a Facebook page, and YouTube videos [73]. When al-Awlaki wanted to marry a white European Muslim convert, Storm stumbled across a Facebook group and found Aminah, a 33-year-old blonde who agreed to become al-Awlaki's third wife [74]. The CIA eventually tracked down al-Awlaki in Yemen and killed him by a drone strike in September 2011 [75].

Whether or not Julian Assange of WikiLeaks is correct in his assertion that "Facebook, Google, and Yahoo have built-in interfaces for the U.S. intelligence to use" [19], law enforcement agencies and the intelligence community have begun to harness the power of social networks in combating international and domestic terrorism.

References

1. **Wright, Lawrence.** The Spymaster. Can Mike McConnell fix America's intelligence community? [Online] The New Yorker, January 21, 2008. http://www.newyorker.com/reporting/2008/01/21/080121fa_fact_wright?currentPage=all.
2. **Niles, Robert.** What's the point of a bag check, anyway? [Online] Theme Park Insider, March 5, 2012. http://www.themeparkinsider.com/flume/201203/2940/.
3. **3VR Inc.** Use Video Analytics and Data Decision Making to Grow Your Business. [Online] Digital Signage Today. [Cited: May 28, 2012.] http://www.digitalsignagetoday.com/whitepapers/4891/Use-Video-Analytics-and-Data-Decision-Making-to-Grow-Your-Business.
4. **Fussey, Pete.** An Interrupted Transmission? Processes of CCTV Implementation and the Impact of Human Agency, Surveillance & Society. [Online] Surveillance and Criminal Justice. 4(3): 229-256, 2007. http://www.surveillance-and-society.org.
5. **Reeve, Tom.** How many cameras in the UK? Only 1.85 million, claims ACPO lead on CCTV. [Online] Security News Desk, March 2011. http://www.securitynewsdesk.com/2011/03/01/how-many-cctv-cameras-in-the-uk/.
6. **Hope, Christopher.** 1,000 CCTV cameras to solve just one crime, Met Police admits. [Online] The Telegraph, August 25, 2009. http://www.telegraph.co.uk/news/uknews/crime/6082530/1000-CCTV-cameras-to-solve-just-one-crime-Met-Police-admits.html.

7. **BBC News.** 'Ring of steel' widened. [Online] BBC News, December 18, 2003. http://news.bbc.co.uk/2/hi/uk_news/england/london/3330771.stm.

8. **Viisage Technology, Inc.** Viisage Technology and Biometrica Systems Achieve 50th Facial Recognition Installation at Mirage Resort, Las Vegas. [Online] PR Newswire, March 29, 2012. http://www.prnewswire.com/news-releases/viisage-technology-and-biometrica-systems-achieve-50th-facial-recognition-installation-at-mirage-resort-las-vegas-73268177.html.

9. **Obama, Barack.** State of the Union 2011: President Obama's Full Speech. [Online] ABC News, January 25, 2011. http://abcnews.go.com/Politics/State_of_the_Union/state-of-the-union-2011-full-transcript/story?id=12759395&page=3.

10. **Scherer, Michael.** Inside the Secret World of the Data Crunchers Who Helped Obama Win. [Online] TIME, November 7, 2012. http://swampland.time.com/2012/11/07/inside-the-secret-world-of-quants-and-data-crunchers-who-helped-obama-win/.

11. **Times, The Washington.** Obama: Be careful what you put on Facebook. [Online] The Washington Times, September 8, 2009. http://www.washingtontimes.com/news/2009/sep/08/obama-advises-caution-what-kids-put-facebook/?page=all.

12. **Etzioni, Amitai.** Despite Facebook, privacy is far from dead. [Online] CNN, May 27, 2012. http://www.cnn.com/2012/05/25/opinion/etzioni-facebook-privacy/index.html.

13. **Ortutay, Barbara.** Facebook tops 1 billion users. [Online] USA Today, October 4, 2012. http://www.usatoday.com/story/tech/2012/10/04/facebook-tops-1-billion-users/1612613/.

14. **Greenfield, Rebecca.** Facebook's New Ad Finds 'Real Human Emotion' in Chairs. [Online] The Atlantic Wire, October 4, 2012. http://www.theatlanticwire.com/technology/2012/10/facebook-aims-real-human-emotion-new-ad/57596/.

15. **Facebook.** The Things That Connect Us. [Online] Facebook, October 4, 2012. http://www.youtube.com/watch?v=c7SjvLceXgU.

16. **Haglund, David.** Facebook's Disingenuous New Ad. [Online] Slate, October 4, 2012. http://www.slate.com/blogs/browbeat/2012/10/04/facebook_chair_ad_why_chairs_explained_video_.html.

17. **Takahashi, Corey.** New Office Chair Promotes Concept of 'Active Sitting'. [Online] The Wall Street Journal, July 21, 2007. http://online.wsj.com/article/SB869433022356380000.html.

18. **Lee, Newton.** Facebook Nation: Total Information Awareness. [Online] Springer, September 15, 2012. http://www.amazon.com/Facebook-Nation-Total-Information-Awareness/dp/1461453070.

19. **RT.** WikiLeaks revelations only tip of iceberg—Assange. [Online] RT News, May 2, 2011. http://www.rt.com/news/wikileaks-revelations-assange-interview/.

20. **Hays, Tom.** NYPD is watching Facebook to Fight Gang Bloodshed. [Online] Associated Press, October 2, 2012. http://bigstory.ap.org/article/nypd-watching-facebook-fight-gang-bloodshed.

21. **Gross, Doug.** Library of Congress to archive your tweets. [Online] CNN, April 14, 2010. http://www.cnn.com/2010/TECH/04/14/library.congress.twitter/index.html.

22. **Gross, Doug.** Library of Congress digs into 170 billion tweets. [Online] CNN, January 7, 2013. http://www.cnn.com/2013/01/07/tech/social-media/library-congress-twitter/index.html.

23. **Hennigan, W. J.** It's a bird! It's a spy! It's both. [Online] Los Angeles Times, February 17, 2011. http://articles.latimes.com/2011/feb/17/business/la-fi-hummingbird-drone-20110217.

24. **CNet Live Blog.** Google's 'next dimension' of Maps event. [Online] CNet, June 6, 2012. http://live.cnet.com/Event/Googles_next_dimension_of_Maps_event?Page=6.

25. **Steinert, Bernd.** Imagery Update: Explore your favorite places in high-resolution. [Online] Google Maps, July 27, 2012. http://google-latlong.blogspot.co.uk/2012/07/imagery-update-explore-your-favorite.html.

26. **Allan, Alasdair.** Got an iPhone or 3G iPad? Apple is recording your moves. [Online] O'Reilly Radar, April 20, 2011. http://radar.oreilly.com/2011/04/apple-location-tracking.html.

27. **Kravets, David.** Researcher's Video Shows Secret Software on Millions of Phones Logging Everything. [Online] Wired, November 29, 2011. http://www.wired.com/threatlevel/2011/11/secret-software-logging-video/.

28. **Eckhart, Trevor.** Carrier IQ Part #2. [Online] YouTube, November 28, 2011. http://www.youtube.com/watch?v=T17XQI_AYNo.

29. **Mayer, Jonathan.** Web Policy. Safari Trackers. [Online] Web Policy Blog, February 17, 2012. http://webpolicy.org/2012/02/17/safari-trackers/.

30. **Angwin, Julia and Valentino-Devries, Jennifer.** Google's iPhone Tracking. Web Giant, Others Bypassed Apple Browser Settings for Guarding Privacy. [Online] The Wall Street Journal, February 17, 2012. http://online.wsj.com/article_email/SB10001424052 9702048804045772253804565991766-lMyQjAxMTAyMDEwNjExNDYyWj.html.

31. **Sarno, David.** Twitter stores full iPhone contact list for 18 months, after scan. [Online] Los Angeles Times, February 14, 2012. http://www.latimes.com/business/technology/la-fi-tn-twitter-contacts-20120214,0,5579919.story.

32. **Koebler, Jason.** Police to Use Drones for Spying on Citizens. [Online] US News and World Report, August 13, 2012. http://www.usnews.com/news/articles/2012/08/23/docs-law-enforcement-agencies-plan-to-use-domestic-drones-for-surveillance.

33. **Ahlers, Mike M.** Congress likes drones, but now looks at flip side of use. [Online] CNN, October 26, 2012. http://www.cnn.com/2012/10/25/us/drones-privacy/index.html.

34. Occupy Wall Street's New Drone: 'The Occucopter'. [Online] Time Magazine, December 21, 2011. http://techland.time.com/2011/12/21/occupy-wall-streets-new-drone-the-occucopter/.

35. **Hargreaves, Steve.** Drones go mainstream. [Online] CNNMoney, January 9, 2013. http://money.cnn.com/2013/01/09/technology/drones/index.html.

36. **Farber, Dan.** Google takes Street View off-road with backpack rig. [Online] CNet, June 6, 2012. http://news.cnet.com/8301-1023_3-57448293-93/google-takes-street-view-off-road-with-backpack-rig/.

37. **Stone, Brad.** Google Says It Inadvertently Collected Personal Data. [Online] The New York Times, May 14, 2010. http://bits.blogs.nytimes.com/2010/05/14/google-admits-to-snooping-on-personal-data/.

38. **Landis, Marina.** Google admits to accidentally collecting e-mails, URLs, passwords. [Online] CNN, October 22, 2010. http://articles.cnn.com/2010-10-22/tech/google.privacy.controls_1_wifi-data-alan-eustace-google-s-street-view?_s=PM:TECH.

39. **Spitzer, Ulf.** Making Google Maps more comprehensive with biggest Street View update ever. [Online] Google Maps, October 11, 2012. http://google-latlong.blogspot.co.uk/2012/10/making-google-maps-more-comprehensive.html.

40. **Kelly, Heather.** Is the government doing enough to protect us online? [Online] CNN, July 31, 2012. http://www.cnn.com/2012/07/25/tech/regulating-cybersecurity/index.html.

41. **Mueller, Robert S. III.** Combating Threats in the Cyber World: Outsmarting Terrorists, Hackers, and Spies. [Online] Federal Bureau of Investigation, March 1, 2012. http://www.fbi.gov/news/speeches/combating-threats-in-the-cyber-world-outsmarting-terrorists-hackers-and-spies.

42. **Jackson, Nicholas.** 15 High-Profile Sites That Google Doesn't Want You to See. [Online] The Atlantic, June 21, 2011. http://www.theatlantic.com/technology/archive/2011/06/15-high-profile-sites-that-google-doesnt-want-you-to-see/240766/.

43. **Cashmore, Pete.** Why 2012, despite privacy fears, isn't like Orwell's 1984. [Online] CNN, January 23, 2012. http://www.cnn.com/2012/01/23/tech/social-media/web-1984-orwell-cashmore/index.html.

44. **Mount, Mike.** Navy SEALs punished for revealing secrets to video game maker. [Online] CNN, November 9, 2012. http://security.blogs.cnn.com/2012/11/09/navy-seals-busted-for-giving-secrets-to-make-video-game-more-real/.

45. **Sanger, David E.** Obama Order Sped Up Wave of Cyberattacks Against Iran. [Online] The New York Times, June 1, 2012. http://www.nytimes.com/2012/06/01/world/middleeast/obama-ordered-wave-of-cyberattacks-against-iran.html?pagewanted=all.

46. **Schmitt, Eric.** U.S. to Step Up Drone Strikes Inside Yemen. [Online] The New York Times, April 25, 2012. http://www.nytimes.com/2012/04/26/world/middleeast/us-to-step-up-drone-strikes-inside-yemen.html.

47. **Cloud, David S. and Rodriguez, Alex.** CIA gets nod to step up drone strikes in Pakistan. [Online] Los Angeles Times, June 8, 2012. http://articles.latimes.com/2012/jun/08/world/la-fg-pakistan-drone-surge-20120608.

48. **Shane, Scott and Schmitt, Eric.** Double Agent Disrupted Bombing Plot, U.S. Says. [Online] The New York Times, May 8, 2012. http://www.nytimes.com/2012/05/09/world/middleeast/suicide-mission-volunteer-was-double-agent-officials-say.html?pagewanted=all.
49. **Levine, Adam.** Obama admits to Pakistan drone strikes. [Online] CNN, January 30, 2012. http://security.blogs.cnn.com/2012/01/30/obama-admits-to-pakistan-drone-strikes/.
50. **McConnell, Dugald and Todd, Brian.** Former special forces officers slam Obama over leaks on bin Laden killing. [Online] CNN, August 17, 2012. http://www.cnn.com/2012/08/16/politics/former-seals-obama/index.html.
51. **Petraeus, David H.** Statement to Employees by Director of the Central Intelligence Agency David H. Petraeus on Safeguarding our Secrets. [Online] Central Intelligence Agency, January 23, 2012. https://www.cia.gov/news-information/press-releases-statements/2012-press-releasese-statements/safeguarding-our-secrets.html.
52. **Blanton, Thomas.** Hearing on the Espionage Act and the Legal and Constitutional Implications of Wikileaks. [Online] Committee on the Judiciary, U.S. House of Representatives, December 16, 2010. http://www.gwu.edu/~nsarchiv/news/20101216/Blanton101216.pdf.
53. **RT.** WikiLeaks revelations only tip of iceberg—Assange. [Online] RT, May 3, 2011. http://rt.com/news/wikileaks-revelations-assange-interview/.
54. **Keller, Bill.** WikiLeaks, a Postscript. [Online] The New York Times, February 19, 2012. http://www.nytimes.com/2012/02/20/opinion/keller-wikileaks-a-postscript.html.
55. **Atkins, Christopher R.** "I solemnly swear to defend the Constitution of the United States of America against all enemies, foreign and domestic." [Online] Michael Moore, January 28, 2005. http://www.michaelmoore.com/words/soldiers-letters/i-solemnly-swear-to-defend-the-constitution-of-the-united-states-of-america-against-all-enemies-foreign-and-domestic.
56. **Wales, Jimmy.** An appeal from Wikipedia founder Jimmy Wales. [Online] Wikimedia Foundation, October 30, 2010. http://wikimediafoundation.org/wiki/Appeal2/en.
57. **Solomon, Brett.** The U.N. Shouldn't Make Decisions About an Open Internet Behind Closed Doors. [Online] Wired, November 30, 2012. http://www.wired.com/opinion/2012/11/you-cant-make-decisions-about-the-open-internet-behind-closed-doors/.
58. **Nxt.** Why we are making all WCIT documents public. [Online]. Nxt, November 22, 2012. http://news.dot-nxt.com/2012/11/23/why-we-are-making-all-wcit-doc.
59. **Kravets, David.** Internet hangs in balance as world leaders meet in secret. [Online] Wired, December 3, 2012. http://www.cnn.com/2012/12/03/tech/web/world-conference-international-telecommunications/index.html.
60. **Cowie, James.** Egypt Leaves the Internet. [Online] Renesys Blog, January 27, 2011. http://www.renesys.com/blog/2011/01/egypt-leaves-the-internet.shtml.
61. **AFP.** Google unveils Web-free 'tweeting' in Egypt move. [Online] Google, January 31, 2011. http://www.google.com/hostednews/afp/article/ALeqM5h8de3cQ8o_S2zg9s72t7sxNToBqA?docId=CNG.ddc0305146893ec9e9e6796d743e6af7.c81.
62. **The Associated Press.** Internet down nationwide in Syria. [Online] USA Today, November 30, 2012. http://www.usatoday.com/story/news/world/2012/11/29/internet-syria/1735721/.
63. **The Economist.** Caught in the net. [Online] The Economist, January 23, 2003. http://www.economist.com/node/1534249.
64. **Paulson, Matthew.** Politics of the Future—How the Internet is Changing and Will Change Politics Forever. [Online] Yahoo! News, November 6, 2006. http://voices.yahoo.com/politics-future-internet-changing-105886.html.
65. **Fitzpatrick, Alex.** U.S. refuses to sign UN Internet treaty. [Online] CNN, December 14, 2012. http://www.cnn.com/2012/12/14/tech/web/un-internet-treaty/index.html.
66. **ONI Team.** Global Internet filtering in 2012 at a glance. [Online] OpenNet Initiative, April 3, 2012. http://opennet.net/blog/2012/04/global-internet-filtering-2012-glance.
67. Internet Freedom. [Online] Freedom House. [Cited: December 4, 2012.] http://www.freedomhouse.org/issues/internet-freedom.
68. **Cerf, Vinton.** 'Father of the internet': Why we must fight for its freedom. [Online] CNN, November 30, 2012. http://edition.cnn.com/2012/11/29/business/opinion-cerf-google-internet-freedom/index.html.

69. **Central Intelligence Agency.** Intellipedia Celebrates Third Anniversary with a Successful Challenge. [Online] Central Intelligence Agency, April 29, 2009. https://www.cia.gov/news-information/featured-story-archive/intellipedia-celebrates-third-anniversary.html.

70. **Bain, Ben.** A-Space set to launch this month. [Online] Federal Computer Week, September 3, 2008. http://fcw.com/articles/2008/09/03/aspace-set-to-launch-this-month.aspx.

71. **Yasin, Rutrell.** National security and social networking are compatible. [Online] Government Computer News, July 23, 2009. http://gcn.com/Articles/2009/07/23/Social-networking-media-national-security.aspx?Page=3.

72. **Bamford, James.** The NSA Is Building the Country's Biggest Spy Center (Watch What You Say). [Online] Wired, March 15, 2012. http://www.wired.com/threatlevel/2012/03/ff_nsadatacenter/all/.

73. **Madhani, Aamer.** Cleric al-Awlaki dubbed 'bin Laden of the Internet'. [Online] USA Today, August 24, 2010. http://usatoday30.usatoday.com/news/nation/2010-08-25-1A_Awlaki25_CV_N.htm.

74. **Cruickshank, Paul, Lister, Tim and Robertson, Nic.** The Danish agent, the Croatian blonde and the CIA plot to get al-Awlaki. [Online] CNN, October 16, 2012. http://www.cnn.com/2012/10/15/world/al-qaeda-cia-marriage-plot/index.html.

75. **Mazzetti, Mark, Schmitt, Eric and Worth, Robert F.** Two-Year Manhunt Led to Killing of Awlaki in Yemen. [Online] The New York Times, September 30, 2011. http://www.nytimes.com/2011/10/01/world/middleeast/anwar-al-awlaki-is-killed in yemen.html?pagewanted=all.

Chapter 7
Cyber Warfare: Weapon of Mass Disruption

> *This world—cyberspace—is a world that we depend on every single day.... America's economic prosperity in the 21st century will depend on cyber security.*
> —President Barack Obama's remark from the White House (May 29, 2009).

> *Terrorism does remain the FBI's top priority, but in the not too-distant-future we anticipate that the cyber threat will pose the greatest threat to our country.*
> —FBI Director Robert Mueller (March 1, 2012).

> *The war is being fought on three fronts. The first is physical, the second is the world of social networks, and the third is cyber attacks.*
> —Carmela Avner, Israel's Chief Information Officer (November 18, 2012).

7.1 Weapon of Mass Disruption

Like counterterrorism, cyber security is in the forefront of the U.S. national security agenda. President Barack Obama remarked from the White House on May 29, 2009 about "a weapon of mass disruption" [1]:

> We meet today at a transformational moment—a moment in history when our interconnected world presents us, at once, with great promise but also great peril. ... This world —cyberspace—is a world that we depend on every single day. It's our hardware and our software, our desktops and laptops and cell phones and Blackberries that have become woven into every aspect of our lives. It's the broadband networks beneath us and the wireless signals around us, the local networks in our schools and hospitals and businesses, and the massive grids that power our nation. It's the classified military and intelligence networks that keep us safe, and the World Wide Web that has made us more interconnected than at any time in human history.

N. Lee, *Counterterrorism and Cybersecurity*, DOI: 10.1007/978-1-4614-7205-6_7,
© Springer Science+Business Media New York 2013

It's the great irony of our Information Age—the very technologies that empower us to create and to build also empower those who would disrupt and destroy. ... Al-Qaeda and other terrorist groups have spoken of their desire to unleash a cyber attack on our country—attacks that are harder to detect and harder to defend against. Indeed, in today's world, acts of terror could come not only from a few extremists in suicide vests but from a few key strokes on the computer—a weapon of mass disruption.

Federal Bureau of Investigation (FBI) Director Robert Mueller spoke at the 2012 RSA Conference in San Francisco: "In one hacker recruiting video, a terrorist proclaims that cyber warfare will be the warfare of the future. ... Terrorism remains the FBI's top priority. But in the not too distant future, we anticipate that the cyber threat will pose the number one threat to our country. We need to take lessons learned from fighting terrorism and apply them to cyber crime" [2].

A year before in March 2011, computer and network security firm RSA disclosed a massive data breach due to "an extremely sophisticated cyber attack" on its computer systems, compromising the effectiveness of its SecurID system that is being used by more than 25,000 corporations and 40 million users around the world [3]. RSA's executive chairman Art Coviello described the attack as an "advanced persistent threat" (APT) from cyber attackers who were skilled, motivated, organized, and well-funded.

RSA was not the only victim. In 2011, more than 760 organizations including almost 20 % of the Fortune 100 companies had their computer networks compromised by some of the same resources used to hit RSA [4]. There were financial firms (e.g. Charles Schwab, Freddie Mac, PriceWaterhouseCoopers, Wells Fargo Bank, World Bank), technology companies (e.g. Amazon.com, AT&T, Cisco, eBay, Facebook, Google, IBM, Intel, Motorola, Microsoft, Sprint, Verizon, Yahoo!), governments (e.g. U.S. Internal Revenue Service, Singapore Government Network), and universities (e.g. MIT, Princeton University, University of California, University of Virginia).

Though not as terrifying as weapons of mass destruction, cyber attacks can be powerful weapons of mass disruption.

7.2 Financial Disruption

President Barack Obama said in his May 2009 remark on "Securing our Nation's Cyber Infrastructure" [1]:

It's about the privacy and the economic security of American families. We rely on the Internet to pay our bills, to bank, to shop, to file our taxes. But we've had to learn a whole new vocabulary just to stay ahead of the cyber criminals who would do us harm—spyware and malware and spoofing and phishing and botnets. Millions of Americans have been victimized, their privacy violated, their identities stolen, their lives upended, and their wallets emptied. According to one survey, in the past two years alone cyber crime has cost Americans more than $8 billion.

In the aftermath of the 9/11 attacks, a large-scale cyber attack occurred seven days later on September 18, 2001 [5]. Nimda (admin spelled backwards) became the Internet's most widespread computer malware to-date and caught many businesses off guard. "A lot of CEOs got really pissed because they thought they had spent a lot of time and money doing cyber security for the company and—bang!—they got hammered, knocked offline, their records got destroyed, and it cost millions of dollars per company," said Richard Clarke, special adviser to the president for cyberspace security and terrorism czar to President Bill Clinton [6].

Nimda blends all three major methods to infect and disable computers: Worm, virus, and Trojan horse. A worm propagates across a network and reproduces itself without user interaction; a virus incorporates itself into other programs when the virus code is executed by a computer user; and a Trojan horse contains hidden code that performs malicious actions. Nimda spread through email attachments, browsing of compromised web sites, and exploitation of vulnerabilities in the Microsoft IIS web server. Microsoft senior.Net developer Michael Lane Thomas characterized the perpetrators behind Nimda as "industrial terrorists" [7].

Computer Economics put the financial damage caused by Nimda to be $635 million [8]. Nimda, Code Red, SirCam, and other cyber attacks in 2001 accounted for a total loss of $13.2 billion in businesses worldwide. Figure 7.1 shows the 12 costliest computer viruses and worms ever [9, 10]:

Computer viruses and worms often grab media headlines because of the widespread infection of computers globally, resulting in potentially high cumulative economic damages. Nevertheless, the lesser known cyber attacks that target specific companies are just as important. In 1994–1995, the "Phone masters" were accountable for $1.85 million in business losses by hacking into the telephone networks of AT&T, British Telecommunications, GTE, MCI, Southwestern Bell, and Sprint; accessing the credit-reporting databases at Equifax and TRW; and entering the systems of Dun and Bradstreet [23]. In 1995, Russian mathematician Vladimir Levin was convicted of stealing $2.8 million from Citibank in a series of electronic break-ins [24]. In 2000, 16-year-old Canadian teenager "Mafiaboy" launched a Distributed Denial of Service (DDoS) attack that took down Amazon.com, Buy.com, CNN, eBay, E*Trade, Excite, Yahoo!, and ZDNet.com, resulting in $1.7 billion in business losses [25].

Fig. 7.1 The 12 costliest computer viruses and worms

Computer Viruses and Worms	Release Date	Financial Damage
1. MyDoom(11)	2004	$38.5 billion
2. SoBig(12)	2003	$37.1 billion
3. ILOVEYOU(13)	2000	$15 billion
4. Conficker(14)	2007	$9.1 billion
5. Code Red(15)	2001	$2 billion
6. Melissa(16)	1999	$1.2 billion
7. SirCam(17)	2001	$1 billion
8. SQL Slammer(18)	2003	$750 million
9. Nimda(19)	2001	$635 million
10. Sasser(20)	2004	$500 million
11. Blaster(21)	2003	$320 million
12. Morris(22)	1988	$10 million

Aside from intentional cyber attacks, there has been unintentional mayhem in the information age. Despite the absence of malicious intent, software bugs can have devastating effects in the financial world. First, there was the 2010 Flash Crash: On the afternoon of May 6, 2010, the Dow Jones Industrial Average (DJIA) plunged almost 1,000 points due to a technical glitch that triggered erroneous trading in Procter & Gamble and several other NYSE stocks [26]. In August 2012, a software bug on Knight Capital's computers executed a series of automatic orders within an hour, resulting in a loss of $440 million and bringing the company to the edge of bankruptcy [27]. Then there was the infamous Facebook IPO snafu caused by a NASDAQ computer glitch on May 18, 2012 [28]. Swiss global financial services company UBS lost $356 million on Facebook as a result [29]. Similarly, in August 2012, a computer system error halted derivatives trading on the Tokyo Stock Exchange for 95 minutes [30] and knocked the Spanish stock exchange IBEX 35 offline for 5 hours [31].

CNN reported in August 2012 that "stock markets have become increasingly vulnerable to bugs over the last decade thanks to financial firms' growing reliance on high-speed computerized trading. Because the trading is automated, there's nobody to apply the brakes if things go wrong" [27]. Overreliance on computerized automated trading plays into the hands of industrial terrorists.

Global security advisor and futurist Marc Goodman spoke at TEDGlobal 2012 on the technological future of crime. He said, "In the last couple hundred years, we've gone from one person robbing another to train robberies, where one gang could rob 200 people at a time. Now, that's scaled to the Sony PlayStation hack, which affected 100 million people. When in history was it ever possible for one person to rob 100 million?" [32].

Symantec released its "Cybercrime Report in 2011" showing that the net cost of cybercrime totaled $139.6 billion in the United States and $388 billion worldwide [33]. In December 2012, McAfee Labs published a report entitled "Analyzing Project Blitzkrieg, a Credible Threat" that warns of an impending cyber attack on 30 U.S. banks in the Spring 2013 [34]. Previously in October 2012, RSA identified the malware as Gozi Prinimalka Trojan that lies dormant in the infested computers until the prescheduled D-day to launch its attack spree [35].

7.3 Infrastructure Disruption

Army Gen. Keith Alexander, director of the National Security Agency (NSA) and Central Security Service (CSS), is also the commander of U.S. Cyber Command that directs the operation and defense of the military's information networks [36]. At the 2012 Aspen Security Forum in Colorado, Alexander said that there had been a 17-fold increase in computer attacks on American infrastructure between 2009 and 2011; and on a scale of 1–10, the U.S. scored a 3 for preventing a serious cyber attack on a critical part of its infrastructure [37].

Paul Stockton, assistant secretary for Homeland Defense and Americas' Security Affairs (DH&ASA), spoke candidly at the 2012 Aspen Security Forum:

"Our adversaries, state and non-state, are not stupid. They are clever and adaptive. There is a risk that they will adopt a profoundly asymmetric strategy, reach around and attack us here at home, the critical infrastructure that is not owned by the Department of Defense" [38].

In November 2012, the National Research Council of the National Academies released the previously classified 2007 report "Terrorism and the Electric Power Delivery System," warning that the U.S. electric power grid is "inherently vulnerable" to terrorist attacks [39]:

> The U.S. electric power delivery system is vulnerable to terrorist attacks that could cause much more damage to the system than natural disasters such as Hurricane Sandy, blacking out large regions of the country for weeks or months and costing many billions of dollars.
> The power grid is inherently vulnerable physically because it is spread across hundreds of miles, and many key facilities are unguarded. ... Many parts of the bulk high-voltage system are heavily stressed, leaving it especially at risk to multiple failures following an attack. Important pieces of equipment are decades old and lack improved technology for sensing and control that could help limit outages and their consequences— not only those caused by a terrorist attack but also in the event of natural disasters.
> There are also critical systems—communications, sensors, and controls that are potentially vulnerable to cyber attacks, whether through Internet connections or by direct penetration at remote sites. Any telecommunication link that is even partially outside the control of the system operators could be an insecure pathway into operations and a threat to the grid.

The August 2003 Northeast blackout deprived 50 million people of electrical power [40]. Many lost water service, gas stations were forced to shut down, and the New York Police Department responded as if the massive power failure were the result of a terrorist attack. It turned out that a software bug in an electricity monitoring system failed to notify operators about a local power outage which then triggered a domino effect, cutting off power in eight U.S. states and parts of Ontario. The total impact on U.S. workers, consumers, and taxpayers was a loss of $6.4 billion [41].

"Power system disruptions experienced to date in the United States, be they from natural disasters or malfunctions, have had immense economic impacts," said Prof. M. Granger Morgan of Carnegie Mellon University and chair of the committee that wrote the 2007 classified report. "Considering that a systematically designed and executed terrorist attack could cause disruptions even more widespread and of longer duration, it is no stretch of the imagination to think that such attacks could produce damage costing hundreds of billions of dollars" [39]. Morgan urged the U.S. government and the power industry to invest in power system research and to develop recovery high-voltage transformers.

Companies are increasingly using the Internet to manage electric power generation, oil-pipeline flow, and water levels in dams remotely by means of "Supervisory Control and Data Acquisition" (SCADA) systems. A SCADA is consisted of a central host that monitors and controls smaller Remote Terminal Units (RTUs) throughout a plant or in the field. The U.S. National Communications System (NCS) released a technical information bulletin in October 2004 indicating that SCADA is susceptible to the various attacks [42]:

- Use a Denial of Service (DoS) attack to crash the SCADA server leading to shutdown condition (System Downtime and Loss of Operations)

- Delete system files on the SCADA server (System Downtime and Loss of Operations)
- Plant a Trojan and take complete control of system (Gain complete control of system and be able to issue any commands available to Operators)
- Log keystrokes from Operators and obtain usernames and passwords (Preparation for future take down)
- Log any company-sensitive operational data for personal or competition usage (Loss of Corporate Competitive Advantage)
- Change data points or deceive Operators into thinking control process is out of control and must be shut down (Downtime and Loss of Corporate Data)
- Modify any logged data in remote database system (Loss of Corporate Data)
- Use SCADA Server as a launching point to defame and compromise other system components within corporate network (IP Spoofing).

In January 2003, the SQL Slammer worm (see Fig. 7.1) penetrated a private computer network protected by a firewall at the Davis-Besse nuclear power plant in Oak Harbor, Ohio [43]. The worm disabled the Safety Parameter Display System (SPDS) and crashed the Plant Process Computer (PPC) for almost five hours. SPDS monitors the nuclear power plant's most crucial safety indicators such as coolant systems, core temperature sensors, and external radiation sensors, whereas PPC monitors the less critical components of the nuclear power plant. An energy sector cyber security expert spoke on condition of anonymity, "If a non-intelligent worm can get in, imagine what an intruder can do" [43].

Indeed, it happened less than seven years later. In June 2010, the Stuxnet worm was discovered by cyber security experts at VirusBlokAda in Belarus. Stuxnet has "all the hallmarks of weaponized software" and it was written specifically to exploit vulnerabilities in SCADA systems that control critical equipment at electric power companies, manufacturing facilities, water treatment plants, nuclear power stations, and other industrial operations [44].

Stuxnet is highly infectious and difficult to detect. It propagates via USB storage devices, injects code into system processes, and hides itself [45]. Roel Schouwenberg, a senior antivirus researcher at Kaspersky Lab, called the Stuxnet worm a "groundbreaking" piece of malware that exploited not just one but four zero-day Windows bugs: Windows shortcuts bug, print spooler bug, and two elevation of privilege (EoP) bugs [46]. Moreover, the malware employed at least two signed digital certificates to make it look legitimate.

Before Microsoft could issue patches to fix the four Windows bugs, Stuxent had infected more than 14,000 computers—nearly 60 % of which were located in Iran, 18 % in Indonesia, 8 % in India, and less than 2 % in Azerbauan, the United States, and Pakistan [47]. The malware was programmed to specifically target SCADA systems manufactured by Siemens [48]. It wreaked havoc on Iran's nuclear facilities and destroyed nearly 1,000 or 20 % of their nuclear centrifuges [49].

David E. Sanger, chief Washington correspondent of *The New York Times*, revealed in June 2012 that President Barack Obama secretly ordered the cyber attacks on Iran's nuclear enrichment facilities by accelerating Operation Olympic

Games that began under the Bush administration in 2006 [50]. A part of the Olympic Games, the Stuxnet worm was a cyber weapon against Iran. However, an error in the program code allowed the worm to escape Iran's nuclear facilities, infect the rest of the world, and inadvertently expose Operation Olympic Games.

Since the world's most sophisticated malware, Stuxnet, has been meticulously dissected and analyzed by cyber security experts worldwide, the concern is that the worm can be reverse-engineered and used against American targets. "Previous cyber attacks had effects limited to other computers," said Michael Hayden, former NSA director and former CIA chief. "This is the first attack of a major nature in which a cyber attack was used to effect physical destruction. Somebody crossed the Rubicon" [50].

Between 2009 and 2011, the number of cyber attacks on infrastructure reported to the U.S. Department of Homeland Security's (DHS) Industrial Control Systems Cyber Emergency Response Team (ICS-CERT) grew by a whopping 2,200 %, from nine to 198 [51]. Attacks against the energy sector represented 41 % of the total number of incidents whereas the water sector had the second highest number of attacks, representing 15 % of all incidents. Figure 7.2 shows that buffer overflow remains the most common vulnerability type among 171 unique vulnerabilities affecting Industrial Control Systems (ICS) products in fiscal year 2012.

In March 2012, ICS-CERT identified an active series of cyber intrusions targeting natural gas pipeline sector companies [52]. There are approximately 7,200 Internet facing control system devices across the United States. (See Fig. 7.3). It is more urgent now than ever before to secure the U.S. infrastructure against cyber attacks.

Fig. 7.2 Vulnerability types affecting Industrial Control Systems in FY 2012 (Courtesy of the Department of Homeland Security ICS-CERT)

Buffer Overflow	44
Input Validation	13
Resource Exhaustion	8
Authentication	8
Cross-site Scripting	8
Path Traversal	8
Resource Management	8
Access Control	7
Hard-coded Password	7
DLL Hijacking	6
SQL Injection	4
Credentials Management	3
Cryptographic Issues	3
Insufficient Entropy	3
Use After Free	3
Use of Hard-coded Credentials	2
Cross-Site Request Forgery	2
Privilege Management	2
Write-what-where Condition	2
Integer Overflow or Wraparound	2
Inadequate Encryption Strength	2
Missing Encryption of Sensitive Data	1
Code Injection	1
Forced Browsing	1
Miscellaneous	15
Total	**171**

Fig. 7.3 7,200 Internet facing control system devices in the U.S. in FY 2012 (Courtesy of the Department of Homeland Security ICS-CERT)

U.S. Department of Homeland Security advocates a "defense-in-depth" approach to securing operations in vital sectors such as electricity, oil and gas, water, transportation, and chemical [53]. ICS-CERT recommends the following industry best practices:

- Minimize network exposure for all control system devices. Control system devices should not directly face the Internet.
- Locate control system networks and devices behind firewalls, and isolate them from the business network.
- If remote access is required, employ secure methods, such as Virtual Private Networks (VPNs), recognizing that VPN is only as secure as the connected devices.
- Remove, disable, or rename any default system accounts wherever possible.
- Implement account lockout policies to reduce the risk from brute forcing attempts.
- Implement policies requiring the use of strong passwords.
- Monitor the creation of administrator level accounts by third-party vendors.
- Adopt a regular patch life cycle to ensure that the most recent security updates are installed.

7.4 Government and Military Disruption

Hackers were trying to tamper with Obama's 2008 presidential campaign, as President Barack Obama explained in his May 2009 remark on "Securing our Nation's Cyber Infrastructure" [1]:

> I know how it feels to have privacy violated because it has happened to me and the people around me. It's no secret that my presidential campaign harnessed the Internet and

technology to transform our politics. What isn't widely known is that during the general election hackers managed to penetrate our computer systems. … Between August and October, hackers gained access to emails and a range of campaign files, from policy position papers to travel plans. And we worked closely with the CIA—with the FBI and the Secret Service and hired security consultants to restore the security of our systems.

In 2008, the U.S. Department of Defense (DoD) suffered a significant compromise of its classified military computer networks. It took the Pentagon nearly 14 months to clean out the computer worm agent.btz [54]. Then U.S. Deputy Secretary of Defense William Lynn wrote, "It began when an infected flash drive was inserted into a U.S. military laptop at a base in the Middle East. The flash drive's malicious computer code, placed there by a foreign intelligence agency, uploaded itself onto a network run by the U.S. Central Command. That code spread undetected on both classified and unclassified systems, establishing what amounted to a digital beachhead, from which data could be transferred to servers under foreign control. It was a network administrator's worst fear: a rogue program operating silently, poised to deliver operational plans into the hands of an unknown adversary" [55].

In September 2011, a keylogger virus penetrated the ground control station (GCS) at the Creech Air Force Base in Nevada and infected the cockpits of the U.S. Predator and Reaper drones [56]. The virus logged pilots' every keystroke as they remotely control the drones in their missions over Afghanistan and other countries. The virus was so resilient that it could not be wiped out by malware removal tools without completely reformatting the GCS' internal hard drives.

In December 2012, the Miami-Dade County Grand Jury reported that "someone created a computer program that automatically, systematically and rapidly submitted to the County's Department of Elections numerous bogus on-line requests for absentee ballots … from a grouping of several different Internet Protocol (IP) addresses … tracked to anonymizers overseas" [57].

In February 2013, personal information of more than 4,000 U.S. bank executives was stolen from the Federal Reserve System by exploiting a temporary vulnerability in a website vendor product [58]. Fortunately, the data breach incident did not affect critical operations of the U.S. central banking system.

John Gilligan, Chief Information Officer of the U.S. Air Force, stated that 80 % of successful penetrations of federal government computers could be attributed to software bugs, trapdoors, and "Easter eggs" often found in commercial off-the-shelf (COTS) products [59].

In early 1980s, the DoD funded the development of a new programming language—Ada—that excels in mission-critical safety and security features [60]. The language was named after mathematician Ada Lovelace (1815–1852) who is often referred to as the world's first programmer. She worked with Charles Babbage, inventor of the first mechanical computer [61]. Ada was meant to be the de facto standard programming language for the U.S. government; and I was involved in drafting the Military Standard Common APSE (Ada Programming Support Environment) Interface Set (CAIS) in 1984 [62]. However, the DoD rescinded its Ada Mandate in 1997 and opted for COTS technology instead [63]. The government's shift from building expensive proprietary software to using cheaper off-the-shelf components shortens development time and hastens deployment; but at the same time the government needs to spend more in testing COTS software programs to look for vulnerabilities.

Indeed, a seemingly harmless and routine Microsoft software update is the masquerade for the Flame virus that had evaded detection for years [64]. A part of Operation Olympic Games, Flame is a data-mining virus that "copies what you enter on your keyboard, monitors what you see on your computer screen, records sounds if the computer is connected to a microphone" [65]. Cyber security firm Kaspersky Lab announced the discovery of the Flame virus in May 2012 [66]:

> Kaspersky Lab's experts, in coordination with ITU (International Telecommunication Union), came across a new type of malware, now known as Flame. Preliminary findings indicate that this malware has been "in the wild" for more than two years—since March 2010. Due to its extreme complexity, plus the targeted nature of the attacks, no security software detected it.

The primary purpose of Flame appears to be cyber espionage, by stealing information from infected machines. Such information is then sent to a network of command-and-control servers located in many different parts of the world. The diverse nature of the stolen information, which can include documents, screenshots, audio recordings and interception of network traffic, makes it one of the most advanced and complete attack-toolkits ever discovered.

The Flame virus targeted at Iran successfully infected the computers of high-ranking Iranian officials [65]. A similar state-sponsored virus, Gauss, was discovered by Kaspersky Lab three months later in August 2012 [67]. Gauss was designed to collect computer information and steal credentials for specific banking, social network, email, and instant messaging accounts. The geographic distribution of Gauss was mostly limited to Lebanon, Israel, and the Palestinian territories, with relatively few infections in the United States and elsewhere in the world.

7.5 Counterfeit Parts and Backdoors

Apart from malware and cyber attacks, counterfeit electronic parts have infiltrated the U.S. government. Lieut. Gen. Patrick J. O'Reilly, director of the Missile Defense Agency (MDA), testified before the U.S. Senate Armed Services Committee in November 2011 [68]:

> MDA has encountered incidents of counterfeit parts dating back to 2006. Total counterfeit parts found to date number about 1,300. All of them were procured from Unauthorized Distributors. We estimate the total cost to MDA for the seven instances is about $4 million. Our largest case cost the Agency $3 million to remove counterfeit parts discovered in the mission computer of our production THAAD (Theater High-Altitude Area Defense) interceptor.
>
> Because counterfeiting continually evolves in sophistication, it is possible that electronic parts may have embedded functionality created by an enemy seeking to disable a system or obtain critical information. Detecting hidden functionality would be a difficult undertaking.
>
> The predominant threat of counterfeit parts in missile defense systems is reduced reliability of a major DoD weapon system. We do not want to be in a position where the reliability of a $12 million THAAD interceptor is destroyed by a $2 part.

O'Reilly said that the MDA had no indication of any mission-critical hardware in the fielded BMDS (Ballistic Missile Defense System) containing counterfeit parts, but he acknowledged that detecting hidden functionality in counterfeit electronic parts poses a significant challenge to the U.S. military. The counterfeit parts can provide backdoors for computer viruses and worms to enter the system during a cyber attack. Backdoors allow cybercriminals to bypass normal authentication processes, firewalls, and other security controls.

Although the U.S. General Services Administration (GSA) has a database of about 90,000 risky suppliers that government agencies are required to check against when ordering electronic parts [69], more than 1,800 instances of counterfeit electronic parts have been found in the defense supply chain to date, and there is no telling of how many more that have not yet been discovered [70].

In 2008, FBI's Operation Cisco Raider resulted in more than 400 seizures of counterfeit Cisco network hardware and labels with an estimated retail value of over $76 million [71]. A number of government agencies bought the routers from an authorized Cisco vendor, but that legitimate vendor purchased the routers from a high-risk Chinese supplier [69]. The Cisco routers manufactured in China could have provided hackers a backdoor into secured U.S. government networks.

U.S. Senator John McCain said in November 2011, "Counterfeit parts pose an increasing risk to our national security, to the reliability of our weapons systems and to the safety of our men and women in uniform" [70].

7.6   Proliferation of Cyber Weapons and Reverse Engineering

Microsoft's principal security architect Roger Grimes wrote in a January 2011 *InfoWorld* article: "With the announcement of the purported success of Stuxnet, the next-generation arms race is on. Ironically, while Stuxnet has possibly slowed down the international proliferation of nuclear arms, it's also officially launched the next big weapons battle" [72].

Eugene Kaspersky, CEO and co-founder of Kaspersky Lab, commented on the discovery of the Flame virus in May 2012: "The risk of cyber warfare has been one of the most serious topics in the field of information security for several years now. Stuxnet and Duqu belonged to a single chain of attacks, which raised cyberwar-related concerns worldwide. The Flame malware looks to be another phase in this war, and it's important to understand that such cyber weapons can easily be used against any country. Unlike with conventional warfare, the more developed countries are actually the most vulnerable in this case" [66].

On May 2, 2011, U.S. Navy SEALs destroyed an inoperable top-secret stealth Blackhawk Helicopter with explosives at the end of the bin Laden raid in Abbottabad, Pakistan. However, the tail section of the craft survived the explosion and the wreckage was hauled away by the Pakistani military. Former White House counterterrorism advisor Richard Clarke commented, "There are probably people in the Pentagon tonight who are very concerned that pieces of the helicopter may be, even now, on their way to China, because we know that China is trying to make stealth aircraft" [73]. It is very likely that other countries have been busy reverse-engineering the stealth technology from the Blackhawk Helicopter wreckage.

One should never downplay the potential of reverse engineering. In 1987 at AT&T Bell Laboratories, I worked with a team of scientists in reverse engineering the Mac OS System 4.1 running on the Apple Macintosh SE. After months of development, we were able to run the Mac OS, HyperCard, and almost all other Macintosh applications on a Unix-based RISC (Reduced Instruction Set Computer) machine. Not only that, but the Mac software ran faster on Unix emulating the Mac OS than on the native Macintosh.

Unless a computer virus or worm is programmed to auto self-destruct when its cover is blown without intervention from the command-and-control (CnC) servers, it is dangerous to unleash highly sophisticated and effective cyber weapons like Stuxnet, Flame, and Gauss. For that reason, some malwares are configured to self-destruct after a certain number of days. The Duqu keylogger worm, for instance, was found to automatically remove itself from an infected system after 36 days of collecting keystrokes and sending them to the perpetrators [74].

7.7 Escalation of Cyber Warfare

In retaliation for the Stuxnet and Flame attacks against Iran's government and nuclear facilities, Iran launched a major cyber attack on American banks in September 2012 using botnets that involved thousands of high-powered application servers. Bank of America, JPMorgan Chase, PNC Bank, U.S. Bank, and Wells Fargo Bank were among the financial institutions that suffered slowdowns and sporadic outages. "The volume of traffic sent to these sites is frankly unprecedented," said cyber security expert Dmitri Alperovitch who conducted the investigation. "It's 10–20 times the volume that we normally see, and twice the previous record for a denial of service attack" [75]. Iran was also believed to be behind the cyber attacks on Saudi Oil company Aramco and on Exxon Mobil's subsidiary RasGas in Qatar [76].

In Syria, supporters of dictator Bashar al-Assad have impersonated opposition activists in order to pass out Trojan horse viruses on Skype calls and via emails [77]. Discovered in February 2012, the cyber espionage malware spies on opposition activists and sends information back to the government-controlled Syrian Telecommunications Establishment (STE).

In November 2012, Israeli Defense Force (IDF) live tweeted its military campaign in the Gaza strip during the weeklong Operation Pillar of Defense [78]. In response, Hamas tweeted its own account of the war along with photographs of casualties [79]. Both sides hoped to use social media to win world sympathy and shift political opinion to their sides [80].

In protest of the Israeli military action in Gaza, hacktivist group Anonymous launched a cyber attack on Israel dubbed "OpIsrael." Anonymous deleted the online databases of the Israel Ministry of Foreign Affairs and Bank of Jerusalem, took down over 663 Israel websites with DDoS attacks, and posted on the Internet over 3,000 emails, phone numbers, and addresses of "Israeli supporters" [81, 82, 83]. Anonymous announced, "November 2012 will be a month to remember for the Israeli defense forces and internet security forces. Israeli Gov. this is/will turn into a cyberwar" [84].

Israel's Chief Information Officer Carmela Avner admitted, "The war is being fought on three fronts. The first is physical, the second is the world of social networks, and the third is cyber attacks" [84]. Indeed, after the ceasefire between Israel and Hamas on November 21, 2012, supporters of both sides intensified their DDoS attacks against each other [85].

In January 2013, the battle for public opinion extended to Hollywood: Iran announced its plan to fund a high-budget movie—*The General Staff*—as a counter-story to the 1979 Iran hostage crisis depicted in the Golden Globe and Academy Award winning film *Argo* [86]. In March 2013, Iran planned to sue Hollywood filmmakers, citing that "the Iranophobic American movie attempts to describe Iranians as overemotional, irrational, insane, and diabolical while at the same, the CIA agents are represented as heroically patriotic" [87].

7.8 Cyber Cold War

While Middle East violence has garnered a lot of media attention, a "cyber Cold War" has been raging under the public radar for over a decade between China and the United States. Prescott Winter, former Chief Technology Officer at the NSA, said, "It's no secret that government agencies are under attack from China. It's a significant problem, and the government has been aware of it for the past 10–15 years" [88].

On April 8, 2010, Chinese hackers "hijacked" the Internet for 18 min by redirecting 15 % of the world's online traffic to route through Chinese servers. The U.S.-China Economic and Security Review Commission released a report in November 2010: "This incident affected traffic to and from U.S. government (".gov") and military (".mil") sites, including those for the Senate, the army, the navy, the marine corps, the air force, the office of secretary of Defense, the National Aeronautics and Space Administration, the Department of Commerce, the National Oceanic and Atmospheric Administration, and many others. Certain commercial websites were also affected, such as those for Dell, Yahoo!, Microsoft, and IBM" [89].

In May 2011, People's Liberation Army (PLA) in China formally announced the deployment of a cyber security squad known as "online blue army" to protect the country from cyber attacks [90]. The Chinese blue army has built the "Great Firewall" (named after the Great Wall of China) to fence off foreign influence by censoring websites and online searches.

In January 2013, *The New York Times* and *The Wall Street Journal* accused Chinese hackers of breaking into their computer systems in order to monitor the news media coverage of China and to look for the names of informants [91]. The Chinese Foreign Ministry vehemently denied the accusations [92].

James A. Lewis, Senior Fellow at the Center for Strategic and International Studies (CSIS), estimated that at least 12 of the world's 15 largest militaries have been building cyber warfare programs [93].

In March 2013, a massive cyber attack against South Korean banks and broadcasters damaged 32,000 computers at their media and financial companies [94]. North Korea, the suspected culprit, has accused South Korea and the United States for daily "intensive and persistent virus attacks" on the country's Internet servers [95]. Since the 9/11 attacks, the NSA oversees "Tailored Access Operations" and the U.S. Navy operates "Computer Network Exploitation" [96]. In selecting a new service motto for the U.S. Air Force in October 2010, Education and Training Command Commander Gen. Stephen Lorenz said, "Airmen consistently told us they see themselves, and they see the heritage of the Air Force, as those entrusted by the nation to defend the modern, complex security domains—first air, then space and now cyberspace" [97]. There is no doubt that cyber war is intensifying and that nations have been stepping up their preparation for state-sponsored cyber attacks.

7.9 Cyber Terrorism and Digital Pearl Harbor

Barry Collin coined the term "cyber terrorism" around 1987 to describe terrorism at the convergence of the physical and virtual worlds [98]. In 2002, James A. Lewis at CSIS defined cyber terrorism as "the use of computer network tools to shut down critical national infrastructures (such as energy, transportation, government operations) or to coerce or intimidate a government or civilian population" [99]. In 2011, retired FBI agent William L. Tafoya refined the definition of cyber terrorism to the criminal acts that target civilians [100]:

> Cyber terrorism is a component of information warfare, but information warfare is not cyber terrorism. … The skills, tools, and techniques are the same, but information warfare is conducted between military combatants; cyber terrorism targets civilians. Cyber terrorists indiscriminately will attack the nation's critical infrastructure and civilians—the innocent. Thus, the context and targets, not the technological tools or frequency of attacks, are the more appropriate delimiters that distinguish cyber terror from information warfare. … Attacking the largely civilian critical infrastructure is not warfare, but terrorism—cyber terror.

Tafoya pointed out that SCADA systems were vulnerable to attacks not only from computer viruses and worms but also from electromagnetic pulse (EMP) bombs and high-energy radio frequency (HERF) weapons. EMP and HERF devices use electromagnetic radiation to deliver heat, mechanical, or electrical energy to an electronic device such as a computer, a cell phone, or even an artificial cardiac pacemaker and defibrillator designed to save lives.

In April 2008, the EMP Commission issued a comprehensive 208-page report on the effects of an EMP attack on U.S. electric power, telecommunications, banking and finance, petroleum and natural gas, transportation, food infrastructure, water works, emergency services, and satellites [101]. The commission cited an incident in November 1999 when San Diego County Water Authority and San Diego Gas and Electric experienced severe electromagnetic interference to their SCADA wireless networks due to a radar operating on a ship 25 miles off the coast of San Diego. Their recommendations are three-pronged: prevention, protection, and recovery.

In October 2012, U.S. Secretary of Defense Leon E. Panetta warned that the United States was facing the possibility of a digital Pearl Harbor: "An aggressor nation or extremist group could use these kinds of cyber tools to gain control of critical switches. They could derail passenger trains, or even more dangerous, derail passenger trains loaded with lethal chemicals. They could contaminate the water supply in major cities, or shut down the power grid across large parts of the country" [102]. Cyber security expert Chiranjeev Bordoloi concurred, "These types of attacks could grow more sophisticated, and the slippery slope could lead to the loss of human life" [103].

The FBI has a division dedicated to combating cyber crime and cyber terrorism. Keith Lourdeau, deputy assistant director of the FBI Cyber Division, testified before the Senate Judiciary Subcommittee on Terrorism, Technology, and Homeland Security in February 2004 [104]:

> The number of individuals and groups with the ability to use computers for illegal, harmful, and possibly devastating purposes is on the rise. We are particularly concerned about terrorists and state actors wishing to exploit vulnerabilities in U.S. systems and networks. … Counterterrorism efforts must incorporate elements from—and contribute toward—counter-intelligence, cyber, and criminal programs.

In December 2012, Dr. Mathew Burrows of the National Intelligence Council (NIC) published a 166-page report, "Global Trends 2030: Alternative Worlds," in which he expressed his outlook on the future of terrorism: "Terrorists for the moment appear focused on causing mass casualties, but this could change as they understand the scope of the disruptions that can be caused by cyber warfare" [105].

References

1. **Obama, Barack.** Remarks By The President On Securing Our Nation's Cyber Infrastructure. [Online] The White House, May 29, 2009. http://www.whitehouse.gov/the-press-office/remarks-president-securing-our-nations-cyber-infrastructure.
2. **Mueller, Robert S. III.** Remarks by Robert S. Mueller, III. [Online] Federal Bureau of Investigation, March 1, 2012. http://www.fbi.gov/news/speeches/combating-threats-in-the-cyber-world-outsmarting-terrorists-hackers-and-spies.
3. **Hickins, Michael and Clark, Don.** Questions Over Break-In at Security Firm RSA. [Online] The Wall Street Journal, March 18, 2011. http://online.wsj.com/article/SB10001424052748703512404576208983743029392.html.
4. **Krebs, Brian.** Who Else Was Hit by the RSA Attackers? [Online] Kerbs on Security, October 11, 2011. http://krebsonsecurity.com/2011/10/who-else-was-hit-by-the-rsa-attackers/.
5. **CERT Coordination Center.** CERT® Advisory CA-2001-26 Nimda Worm. [Online] Carneige Mellon Software Engineering Institute, September 18, 2001. http://www.cert.org/advisories/CA-2001-26.html.
6. **Green, Joshua.** The Myth of Cyberterrorism. [Online] Washington Monthly, November 2002. http://www.washingtonmonthly.com/features/2001/0211.green.html.
7. **Leyden, John.** Virus writers are industrial terrorists–MS. [Online] The Register, October 23, 2001. http://www.theregister.co.uk/2001/10/23/virus_writers_are_industrial_terrorists/.

8. Malicious Code Attacks Had $13.2 Billion Economic Impact in 2001. [Online] Computer Economics, September 2002. http://www.computereconomics.com/article.cfm?id=133.
9. **Marquit, Miranda.** The 12 costliest computer viruses ever. [Online] The Fine Print, August 3, 2010. http://blog.insure.com/2010/08/03/the-12-costliest-computer-viruses-ever/.
10. **McAfee, Inc.** McAfee Looks Back on a Decade of Cybercrime. [Online] McAfee, January 25, 2011. http://www.mcafee.com/us/about/news/2011/q1/20110125-01.aspx.
11. **Help Net Security.** Mydoom.A: Timeline of an Epidemic. [Online] Help Net Security, March 2, 2004. http://www.net-security.org/malware_news.php?id=359.
12. **Roberts, Paul.** Sobig: Spam, virus or both? [Online] Computerworld, June 5, 2003. http://www.computerworld.com/s/article/81825/Sobig_Spam_virus_or_both_.
13. **CNN.** Destructive ILOVEYOU computer virus strikes worldwide. [Online] CNN, May 4, 2000. http://articles.cnn.com/2000-05-04/tech/iloveyou.01_1_melissa-virus-antivirus-companies-iloveyou-virus.
14. **Danchev, Dancho.** Conficker's estimated economic cost? $9.1 billion. [Online] ZDNet, April 23, 2009. http://www.zdnet.com/blog/security/confickers-estimated-economic-cost-9-1-billion/3207.
15. **Reuters.** 'Code Red II' spreading quickly, causing damage. [Online] USA Today, August 8, 2001. http://usatoday30.usatoday.com/life/cyber/tech/2001-08-08-code-red-2.htm.
16. **Northcutt, Stephen.** Intrusion Detection FAQ: What was the Melissa virus and what can we learn from it? [Online] SANS, March 28, 1999. http://www.sans.org/security-resources/idfaq/what_melissa_teaches_us.php.
17. **Thorsberg, Frank.** Sircam Worm: Crawling Fast but Easily Crushed. [Online] PCWorld, July 26, 2001. http://www.pcworld.com/article/56284/article.html.
18. **Lemos, Robert.** Counting the cost of Slammer. [Online] CNet, January 31, 2003. http://news.cnet.com/Counting-the-cost-of-Slammer/2100-1001_3-982955.html.
19. **Lemos, Robert.** 'Nimda' worm strikes Net, e-mail. [Online] CNet, September 18, 2001. http://news.cnet.com/2100-1001-273128.html.
20. **Lemos, Robert.** Sasser worm begins to spread. [Online] CNet, May 1, 2004. http://news.cnet.com/Sasser-worm-begins-to-spread/2100-7349_3-5203764.html.
21. **Messmer, Ellen.** Blaster Worm Racks Up Victims. [Online] PCWorld, August 15, 2003. http://www.pcworld.com/article/112047/article.html.
22. **Markoff, John.** Student, After Delay, Is Charged In Crippling of Computer Network. [Online] The New York Times, July 27, 1989. http://www.nytimes.com/1989/07/27/us/student-after-delay-is-charged-in-crippling-of-computer-network.html.
23. **Simons, John.** How an FBI Cybersleuth Busted a Hacker Ring. [Online] The Wall Street Journal, October 3, 1999. http://massis.lcs.mit.edu/archives/security-fraud/phonemasters-fraud.
24. **Johnston, David Cay.** Russian Accused of Citibank Computer Fraud. [Online] The New York Times, August 18, 1995. http://www.nytimes.com/1995/08/18/business/russian-accused-of-citibank-computer-fraud.html.
25. **Evans, James.** Mafiaboy's Story Points to Net Weaknesses. [Online] PCWorld, January 24, 2001. http://www.pcworld.com/article/39142/article.html.
26. **Twin, Alexandra.** Glitches send Dow on wild ride. [Online] CNNMoney, May 6, 2010. http://money.cnn.com/2010/05/06/markets/markets_newyork/index.htm.
27. **Eha, Brian Patrick.** Is Knight's $440 million glitch the costliest computer bug ever? [Online] CNNMoney, August 9, 2012. http://money.cnn.com/2012/08/09/technology/knight-expensive-computer-bug/index.html.
28. **Yousuf, Hibah.** Facebook trader: Nasdaq 'blew it'. [Online] CNNMoney, May 21, 2012. http://money.cnn.com/2012/05/21/markets/facebook-nasdaq/index.htm.
29. **Yousuf, Hibah.** UBS lost $356 million on Facebook, suing Nasdaq for it. [Online] CNNMoney, July 31, 2012. http://buzz.money.cnn.com/2012/07/31/ubs-loss-facebook-ipo/.
30. **Hasegawa, Toshiro, Nohara, Yoshiaki and Ikeda, Yumi.** Tokyo System Errors Underscore Decline in Japan's Equity Market. [Online] Bloomberg, August 7, 2012. http://www.bloomberg.com/news/2012-08-07/second-system-error-in-seven-months-halts-tokyo-derivative-trade.html.
31. **Rooney, Ben.** Spanish stocks halted for 5 hours due to trading glitch. [Online] CNNMoney, August 6, 2012. http://buzz.money.cnn.com/2012/08/06/spain-stocks-trading-glitch/.

32. **TED.** The technological future of crime: Marc Goodman at TEDGlobal 2012. [Online] TED, June 28, 2012. http://blog.ted.com/2012/06/28/the-technological-future-of-crime-marc-goodman-at-tedglobal-2012/.

33. **Symantec.** Cybercrime Report 2011. [Online] Symantec Corporation, 2011. http://www.symantec.com/content/en/us/home_homeoffice/html/cybercrimereport/assets/downloads/en-us/NCR-DataSheet.pdf.

34. **Sherstobitoff, Ryan.** Analyzing Project Blitzkrieg, a Credible Threat. [Online] McAfee Labs, December 13, 2012. http://www.mcafee.com/us/resources/white-papers/wp-analyzing-project-blitzkrieg.pdf.

35. **RSA FraudAction Research Labs.** Cyber Gang Seeks Botmasters to Wage Massive Wave of Trojan Attacks Against U.S. Banks. [Online] RSA, October 4, 2012. https://blogs.rsa.com/cyber-gang-seeks-botmasters-to-wage-massive-wave-of-trojan-attacks-against-u-s-banks/.

36. What are the differences between NSA/CSS' and U.S. Cyber Command's roles? [Online] National Security Agency / Central Security Service, January 13, 2011. http://www.nsa.gov/about/faqs/about_nsa.shtml#about10.

37. **Sanger, David E. and Schmitt, Eric.** Rise Is Seen in Cyberattacks Targeting U.S. Infrastructure. [Online] The New York Times, July 26, 2012. http://www.nytimes.com/2012/07/27/us/cyberattacks-are-up-national-security-chief-says.html?_r=0.

38. **Merica, Dan.** DoD official: Vulnerability of U.S. electrical grid is a dire concern. [Online] CNN, July 27, 2012. http://security.blogs.cnn.com/2012/07/27/dod-official-vulnerability-of-u-s-electrical-grid-is-a-dire-concern/.

39. **National Research Council of the National Academies.** Electric Power Grid 'Inherently Vulnerable' to Terrorist Attacks. [Online] The National Academies Press, November 14, 2012. http://www8.nationalacademies.org/onpinews/newsitem.aspx?RecordID=12050.

40. **Barron, James.** The Blackout Of 2003: The Overview; Power Surge Blacks Out Northeast, Hitting Cities In 8 States And Canada; Midday Shutdowns Disrupt Millions. [Online] The New York Times, August 15, 2003. http://www.nytimes.com/2003/08/15/nyregion/blackout-2003-overview-power-surge-blacks-northeast-hitting-cities-8-states.html?pagewanted=all.

41. **Anderson, Patrick L. and Geckil, Ilhan K.** Northeast Blackout Likely to Reduce US Earnings by $6.4 Billion. [Online] Anderson Economic Group (AEG), August 19, 2003. http://www.andersoneconomicgroup.com/Portals/0/upload/Doc544.pdf.

42. **Communication Technologies, Inc.** Supervisory Control and Data Acquisition (SCADA) Systems. [Online] National Communications System, October 2004. http://www.ncs.gov/library/tech_bulletins/2004/tib_04-1.pdf.

43. **Poulsen, Kevin.** Slammer worm crashed Ohio nuke plant network. [Online] Security Focus, August 19, 2003. http://www.securityfocus.com/news/6767.

44. **McMillan, Robert.** New virus targets industrial secrets. [Online] Computerworld, July 17, 2010. http://www.computerworld.com/s/article/9179298/New_virus_targets_industrial_secrets.

45. **Ulasen, Sergey.** Rootkit.TmpHider. [Online] Wilders Security Forums, July 12, 2010. http://www.wilderssecurity.com/showthread.php?t=276994.

46. **Keizer, Gregg.** Is Stuxnet the 'best' malware ever? [Online] Computerworld, September 16, 2010. http://www.computerworld.com/s/article/9185919/Is_Stuxnet_the_best_malware_ever_.

47. **Thakur, Vikram.** W32.Stuxnet — Network Information. [Online] Symantec, July 23, 2010. http://www.symantec.com/connect/blogs/w32stuxnet-network-information.

48. **McMillan, Robert.** Siemens: Stuxnet worm hit industrial systems. [Online] Computerworld, September 14, 2010. http://www.computerworld.com/s/article/9185419/Siemens_Stuxnet_worm_hit_industrial_systems.

49. **Broad, William J., Markoff, John and Sanger, David E.** Israeli Test on Worm Called Crucial in Iran Nuclear Delay. [Online] The New York Times, January 15, 2011. http://www.nytimes.com/2011/01/16/world/middleeast/16stuxnet.html?pagewanted=all.

50. **Sanger, David E.** Obama Order Sped Up Wave of Cyberattacks Against Iran. [Online] The New York Times, June 1, 2012. http://www.nytimes.com/2012/06/01/world/middleeast/obama-ordered-wave-of-cyberattacks-against-iran.html?pagewanted=all.

51. **DHS ICS-CERT.** ICS-CERT Monitor. [Online] U.S. Department of Homeland Security, October/November/December 2012. http://www.us-cert.gov/control_systems/ pdf/ICS-CERT_Monthly_Monitor_Oct-Dec2012.pdf.
52. **ICS-CERT.** ICS-CERT Monthly Monitor. [Online] Industrial Control Systems Cyber Emergency Response Team, April 2012. http://www.us-cert.gov/control_systems/ pdf/ICS-CERT_Monthly_Monitor_Apr2012.pdf.
53. **Control Systems Security Program, National Cyber Security Division.** Recommended Practice: Improving Industrial Control Systems Cybersecurity with Defense-In-Depth Strategies. [Online] U.S. Department of Homeland Security, October 2009. http://www.us-cert.gov/control_systems/practices/documents/Defense_in_Depth_Oct09.pdf.
54. **Shachtman, Noah.** Insiders Doubt 2008 Pentagon Hack Was Foreign Spy Attack (Updated). [Online] Wired, August 25, 2010. http://www.wired.com/dangerroom/2010/08/ insiders-doubt-2008-pentagon-hack-was-foreign-spy-attack/.
55. **Lynn, William J. III.** Defending a New Domain. [Online] Foreign Affairs, September/October 2010. http://www.foreignaffairs.com/articles/66552/william-j-lynn-iii/defending-a-new-domain.
56. **Shachtman, Noah.** Exclusive: Computer Virus Hits U.S. Drone Fleet. [Online] Wired, October 7, 2011. http://www.wired.com/dangerroom/2011/10/virus-hits-drone-fleet/.
57. **Rundle, Katherine Fernandez; Horn, Donl; Dechovitz, Susan Leah.** Final Report Of The Miami-Dade County Grand Jury. [Online] The Circuit Court Of The Eleventh Judicial Circuit Of Florida In And For The County Of Miami-Dade, December 19, 2012. http://www .miamisao.com/publications/grand_jury/2000s/gj2012s.pdf
58. **Bull, Alister and Finkle, Jim.** Fed says internal site breached by hackers, no criti-cal functions affected. [Online] Reuters, February 6, 2013. http://www.reuters.com/ article/2013/02/06/net-us-usa-fed-hackers-idUSBRE91501920130206.
59. **Green, Joshua.** The Myth of Cyberterrorism. [Online] Washington Monthly, November 2002. http://www.washingtonmonthly.com/features/2001/0211.green.html.
60. **Ada Resource Association.** Ada Overview. [Online] Ada Information Clearinghouse. [Cited: January 8, 2013.] http://www.adaic.org/advantages/ada-overview/.
61. **Computer History Museum.** Ada Lovelace. [Online] Computer History Museum. [Cited: January 8, 2013.] http://www.computerhistory.org/babbage/adalovelace/.
62. **Ada Joint Program Office.** Military Standard Common APSE (Ada Programming Support Environment) Interface Set (CAIS). [Online] Defense Technical Information Center, 1985. http://books.google.com/books/about/Military_Standard_Common_APSE_Ada_Progra.htm l?id=EjEYOAAACAAJ.
63. **Keller, John.** DOD officials eye scrapping mandate to use Ada programming. [Online] Military and Aerospace Electronics, May 1, 1997. http://www.militaryaerospace.com/ articles/print/volume-8/issue-5/departments/trends/dod-officials-eye-scrapping-mandate-to-use-ada-programming.html.
64. **Nakashima, Ellen, Miller, Greg and Tate, Julie.** U.S., Israel developed Flame com-puter virus to slow Iranian nuclear efforts, officials say. [Online] The Washington Post, June 19, 2012. http://articles.washingtonpost.com/2012-06-19/world/35460741_ 1_stuxnet-computer-virus-malware.
65. **Erdbrink, Thomas.** Iran Confirms Attack by Virus That Collects Information. [Online] The New York Times, May 29, 2012. http://www.nytimes.com/2012/05/30/world/middleeast/ iran-confirms-cyber-attack-by-new-virus-called-flame.html.
66. **Kaspersky Lab security researchers.** Kaspersky Lab and ITU Research Reveals New Advanced Cyber Threat. [Online] Kaspersky Lab, May 28, 2012. http://www.kasper sky.com/about/news/virus/2012/Kaspersky_Lab_and_ITU_Research_Reveals_New_ Advanced_Cyber_Threat.
67. **Global Research & Analysis Team (GReAT), Kaspersky Lab.** Gauss: Abnormal Distribution. [Online] Kaspersky Lab, August 9, 2012. http://www.securelist.com/en/ analysis/204792238/Gauss_Abnormal_Distribution.
68. **O'Reilly, Patrick J.** Lieutenant General Patrick J. O'Reilly, USA, Director, Missile Defense Agency Before the Senate Armed Services Committee. [Online] U.S. Senate

Armed Services Committee, November 8, 2011. http://www.armed-services.senate.
gov/statemnt/2011/11%20November/OReilly%2011-08-11.pdf.

69. **Goldman, David.** Fake tech gear has infiltrated the U.S. government. [Online] CNNMoney,
November 8, 2012. http://money.cnn.com/2012/11/08/technology/security/counterfeit-tech/
index.html.

70. **Carl Levin - U.S. Senate Newsroom.** Senate Approves Amendment to Strengthen
Protections Against Counterfeit Electronic Parts in Defense Supply System. [Online] U.S.
Senate, November 29, 2011. http://www.levin.senate.gov/newsroom/press/release/senate-
approves-amendment-to-strengthen-protections-against-counterfeit-electronic-parts-in-
defense-supply-system.

71. **U.S. Department of Justice.** Departments of Justice and Homeland Security Announce
International Initiative Against Traffickers in Counterfeit Network Hardware. [Online] U.S.
Department of Justice, February 28, 2008. http://www.justice.gov/opa/pr/2008/February/08_
crm_150.html.

72. **Grimes, Roger A.** Stuxnet marks the start of the next security arms race. [Online]
InfoWorld, January 25, 2011. http://www.infoworld.com/d/security-central/stuxnet-marks-
the-start-the-next-security-arms-race-282.

73. **Ross, Brian, et al.** Top Secret Stealth Helicopter Program Revealed in Osama
Bin Laden Raid: Experts. [Online] ABC Good Morning America, May 4, 2011.
http://abcnews.go.com/Blotter/top-secret-stealth-helicopter-program-revealed-osama bin/
story?id=13530693&page-2.

74. **Zetter, Kim.** Son of Stuxnet Found in the Wild on Systems in Europe. [Online] Wired,
October 18, 2011. http://www.wired.com/threatlevel/2011/10/son-of-stuxnet-in-the-wild/.

75. **Goodman, David.** Major banks hit with biggest cyberattacks in history. [Online] CNNMoney,
September 28, 2012. http://money.cnn.com/2012/09/27/technology/bank-cyberattacks/index.html.

76. **Goldman, David.** The real Iranian threat: Cyberattacks. [Online] CNNMoney, November 5,
2012. http://money.cnn.com/2012/11/05/technology/security/iran-cyberattack/index.html.

77. **Brumfield, Ben.** Computer spyware is newest weapon in Syrian conflict. [Online] CNN,
February 17, 2012. http://www.cnn.com/2012/02/17/tech/web/computer-virus-syria/index.html.

78. **Fung, Brian.** Military Strikes Go Viral: Israel Is Live-Tweeting Its Own Offensive Into
Gaza. [Online] The Atlantic, November 14, 2012. http://www.theatlantic.com/international/
archive/2012/11/military-strikes-go-viral-israel-is-live-tweeting-its-own-offensive-into-
gaza/265227/.

79. **Hachman, Mark.** IDF vs. Hamas War Extends to Social Media. [Online] PC Magazine,
November 16, 2012. http://www.pcmag.com/slideshow/story/305065/idf-vs-hamas-war-
extends-to-social-media.

80. **Sutter, John D.** Will Twitter war become the new norm? [Online] CNN, November 19,
2012. http://www.cnn.com/2012/11/15/tech/social-media/twitter-war-gaza-israel/index.html.

81. **Chan, Casey.** Anonymous Targets Israel by Taking Down Hundreds of Websites and
Leaking Emails and Passwords. [Online] Gizmodo, November 16, 2012. http://gizmodo.
com/5961399/anonymous-destroys-israel-by-taking-down-hundreds-of-websites-and-leak-
ing-emails-and-passwords.

82. **Greenberg, Andy.** Anonymous Hackers Ramp Up Israeli Web Attacks And Data Breaches
As Gaza Conflict Rages. [Online] Forbes, November 19, 2012. http://www.forbes.com/sit
es/andygreenberg/2012/11/19/anonymous-hackers-ramp-up-israeli-web-attacks-and-data-
breaches-as-gaza-conflict-rages-2/.

83. **Osborne, Charlie.** Anonymous takes on Israeli websites, wipes Jerusalem bank.
[Online] ZDNet, November 16, 2012. http://www.zdnet.com/anonymous-takes-on-
israeli-websites-wipes-jerusalem-bank-7000007537/.

84. **Sutter, John D.** Anonymous declares 'cyberwar' on Israel. [Online] CNN, November 20,
2012. http://www.cnn.com/2012/11/19/tech/web/cyber-attack-israel-anonymous/index.html.

85. **McMillian, Robert and Ackerman, Spencer.** Despite Ceasefire, Israel-Gaza War
Continues Online. [Online] Wired, November 28, 2012. http://www.wired.com/dangerr
oom/2012/11/israel-gaza-ddos/.

86. **Mullen, Jethro.** Coming soon: Iran's response to 'Argo'. [Online] CNN, January 14, 2013. http://www.cnn.com/2013/01/14/world/meast/iran-argo-response/index.html.

87. **Mullen, Jethro; Brumfield, Ben.** Iran to add lawsuit over 'Argo' to cinematic response. [Online] CNN, March 14, 2013. http://www.cnn.com/2013/03/13/world/meast/iran-argo-response/index.html

88. **Goldman, David.** China vs. U.S.: The cyber Cold War is raging. [Online] CNNMoney, July 28, 2011. http://money.cnn.com/2011/07/28/technology/government_hackers/index.htm.

89. **Ryan, Jason.** US Government and Military Websites Redirected to Chinese Servers. [Online] ABC News, November 17, 2010. http://abcnews.go.com/Technology/american-government-websites-hijacked-chinese-hackers-massive-april/story?id=12165826.

90. **Beech, Hannah.** Meet China's Newest Soldiers: An Online Blue Army. [Online] Time Magazine, May 27, 2011. http://world.time.com/2011/05/27/meet-chinas-newest-soldiers-an-online-blue-army/.

91. **Mullen, Jethro.** New York Times, Wall Street Journal say Chinese hackers broke into computers. [Online] CNN, January 31, 2013. http://www.cnn.com/2013/01/31/tech/china-nyt-hacking/index.html.

92. **Riley, Charles.** China's military denies hacking allegations. [Online] CNNMoney, February 20, 2013. http://money.cnn.com/2013/02/20/technology/china-cyber-hacking-denial/index.html

93. **Shane, Scott.** Cyberwarfare Emerges From Shadows for Public Discussion by U.S. Officials. [Online] 2012, 26 September. http://www.nytimes.com/2012/09/27/us/us-officials-opening-up-on-cyberwarfare.html?pagewanted=all.

94. **Kwon, K.J.; Mullen, Jethro; Pearson, Michael.** Hacking attack on South Korea traced to Chinese address, officials say. [Online] CNN, March 21, 2013. http://www.cnn.com/2013/03/21/world/asia/south-korea-computer-outage/index.html

95. Botelho, Greg. North Korea says it's the victim of 'intensive' cyberattacks. [Online] CNN, March 15, 2013. http://www.cnn.com/2013/03/14/world/asia/north-korea-cyberattacks/index.html

96. **Kingsbury, Alex.** The Secret History of the National Security Agency. [Online] US News and World Report, June 19, 2009. http://www.usnews.com/opinion/articles/2009/06/19/the-secret-history-of-the-national-security-agency.

97. **Air Force News Service (AFNS).** 'Aim High … Fly-Fight-Win' to be Air Force motto. [Online] U.S. Air Force, October 7, 2010. http://www.af.mil/news/story.asp?id=123225546.

98. **Collin, Barry.** The Future of Cyberterrorism. [Online] Crime & Justice International, March 1997. http://www.cjimagazine.com/archives/cji4c18.html?id=415.

99. Assessing the Risks of Cyber Terrorism, Cyber War and Other Cyber Threats. [Online] Center for Strategic and International Studies (CSIS), December 2002. http://csis.org/files/media/csis/pubs/021101_risks_of_cyberterror.pdf .

100. **Tafoya, William L.** Cyber Terror. [Online] Federal Bureau of Investigation, November 2011. http://www.fbi.gov/stats-services/publications/law-enforcement-bulletin/november-2011/cyber-terror.

101. **Foster, John S., et al.** Report of the Commission to Assess the Threat to the United States from Electromagnetic Pulse (EMP) Attack. [Online] The EMP Commission, April 2008. http://www.empcommission.org/docs/A2473-EMP_Commission-7MB.pdf.

102. **Bumiller, Elisabeth and Shanker, Thom.** Panetta Warns of Dire Threat of Cyberattack on U.S. [Online] The New York Times, October 11, 2012. http://www.nytimes.com/2012/10/12/world/panetta-warns-of-dire-threat-of-cyberattack.html?pagewanted=all.

103. **Goldman, David.** Nations prepare for cyber war. [Online] CNNMoney, January 7, 2013. http://money.cnn.com/2013/01/07/technology/security/cyber-war/index.html.

104. **Lourdeau, Keith.** Keith Lourdeau, Cyber Division, FBI, Before the Senate Judiciary Subcommittee on Terrorism, Technology, and Homeland Security. [Online] 2004, 24 February. http://www.fbi.gov/news/testimony/hearing-on-cyber-terrorism.

105. **National Intelligence Council.** Global Trends 2030: Alternative Worlds. [Online] U.S. National Intelligence Council, December 2012. http://www.dni.gov/files/documents/Global Trends_2030.pdf.

Chapter 8
Cyber Attacks, Prevention, and Countermeasures

There are only two types of companies: those that have been hacked, and those that will be. Even that is merging into one category: those that have been hacked and will be again.
—FBI Director Robert Mueller
RSA conference (March 1, 2012).

The attack surfaces for adversaries to get on the Internet now include all those mobile devices. The mobile security situation lags. It's far behind.
—Army Gen. Keith Alexander, Director of National Security Agency and Commander of U.S. Cyber Command
DEF CON 20 (July 27, 2012).

There is no such thing as 100 % security, on- or offline, but we must strive to strengthen our defenses against those who are constantly working to do us harm.... The alternative could be a digital Pearl Harbor—and another day of infamy.
—U.S. senators Joe Lieberman, Susan Collins and Tom Carper
(July 7, 2011).

8.1 Cybersecurity Acts

In response to the ever-increasing number of cyber attacks on both private companies and the United States government, U.S. Congress has introduced the Cybersecurity Enhancement Act of 2007 [1], the National Commission on American Cybersecurity Act of 2008 [2], the Cybersecurity Act of 2009 [3], the Cybersecurity and American Cyber Competitiveness Act of 2011 [4], and most recently the Cybersecurity Act of 2012 [5].

In July 2011, U.S. senators Joe Lieberman, Susan Collins and Tom Carper wrote in *The Washington Post* in support of their cybersecurity bill: "There is no such thing as 100 % security, on- or offline, but we must strive to strengthen our

N. Lee, *Counterterrorism and Cybersecurity*, DOI: 10.1007/978-1-4614-7205-6_8, defenses against those who are constantly working to do us harm.... The alternative could be a digital Pearl Harbor—and another day of infamy" [6].

U.S. Senate Commerce Committee Chairman Jay Rockefeller said at a Senate Intelligence Committee hearing in January 2012, "The threat posed by cyber attacks is greater than ever, and it's a threat not just to companies like Sony or Google but also to the nation's infrastructure and the government itself. Today's cybercriminals have the ability to interrupt life-sustaining services, cause catastrophic economic damage, or severely degrade the networks our defense and intelligence agencies rely on. Congress needs to act on comprehensive cybersecurity legislation immediately" [7].

However, opponents of the cybersecurity acts view the proposed legislations as digital versions of the Patriot Act of 2001, an unnecessary government intrusion into private businesses reminiscent of the Sarbanes-Oxley Act of 2002, or justification for an overreaching "cyber-industrial complex" akin to the expansive military-industrial complex [8–10].

Businesses caution that the cybersecurity acts could harm U.S. competitiveness in the technology sector, and privacy advocates fear that the proposed bills would give the U.S. government too much authority to examine the content of emails, file transfers, and Web searches. Former National Security Agency (NSA) consultant Ed Giorgio said, "Google has records that could help in a cyber-investigation. We have a saying in this business: 'Privacy and security are a zero-sum game'" [11].

8.2 Cybersecurity Initiatives: CNCI, NICE, Presidential Executive Order

While the cybersecurity acts are embattled in heated debates, President Barack Obama signed an executive order in February 2013 to reduce cyber risk to critical infrastructure and to improve cybersecurity information sharing between the private and public sectors [12]. No one could disagree with the 2009 U.S. Cyberspace Policy Review that identified enhanced information sharing as a key component of effective cybersecurity [13]. Established by President George W. Bush in January 2008 and reconfirmed by President Barack Obama in May 2009, the Comprehensive National Cybersecurity Initiative (CNCI) has launched the following initiatives to assure a trusted and resilient information and communications infrastructure across government agencies [14]:

1. Manage the Federal Enterprise Network as a single network enterprise with Trusted Internet.
2. Deploy an intrusion detection system of sensors across the Federal enterprise.
3. Pursue deployment of intrusion prevention systems across the Federal enterprise.
4. Coordinate and redirect research and development (R&D) efforts.
5. Connect current cyber ops centers to enhance situational awareness.
6. Develop and implement a government-wide cyber counterintelligence (CI) plan.

7. Increase the security of our classified networks.
8. Expand cyber education.
9. Define and develop enduring "leap-ahead" technology, strategies, and programs.
10. Define and develop enduring deterrence strategies and programs.
11. Develop a multi-pronged approach for global supply chain risk management.
12. Define the Federal role for extending cybersecurity into critical infrastructure domains.

Education is one of the key CNCI initiatives, and its scope has expanded from a federal focus to a larger national focus. Starting in April 2010, the National Initiative for Cybersecurity Education (NICE) represents the continual evolution of CNCI, and the National Institute of Standards and Technology (NIST) has assumed the overall coordination role [15].

With the support of CNCI and NICE, the U.S. Department of Defense (DoD) Information Assurance Support Environment (IASE) has launched Cyber Awareness Challenge education programs for DoD and Federal employees [16]. The training programs are unclassified and accessible by the general public. The exercises cover a wide range of computer usage and serve as an excellent basic tool for everyone to learn about cybersecurity (Fig. 8.1).

8.3 National Cyber Security Awareness Month (NCSAM)

President Barack Obama designated October as the National Cyber Security Awareness Month (NCSAM). NCSAM is designed to "engage and educate public and private sector partners through events and initiatives with the goal of raising

Fig. 8.1 Cyber awareness challenge (courtesy of U.S. Department of Defense)

awareness about cybersecurity and increasing the resiliency of the nation in the event of a cyber incident" [17].

October 2012 marked the ninth annual National Cyber Security Awareness Month sponsored by the U.S. Department of Homeland Security (DHS) in cooperation with the National Cyber Security Alliance (NCSA) and the Multi-State Information Sharing and Analysis Center (MS-ISAC).

The DHS encourages the entire American community to ACT—Achieve Cybersecurity Together. It reflects the interconnectedness of the modern world and the responsibility that everyone has in securing cyberspace. DHS oversees Cyber Storm, a biennial simulated cyber attacks exercise to strengthen cyber preparedness in the public and private sectors [18]. Cyber Storm I through IV were held in February 2006, March 2008, September 2010, and from Fall 2011 to 2012.

The NCSA publishes the website www.StaySafeOnline.org to educate computer users how to stay safe online, teach online safety to children and adults, keep business safe online, and get involved in National Cyber Security Awareness Month (NCSAM), Data Privacy Day (DPD), and National Cyber Security Education Council (NCEC) [19].

The MS-ISAC is a 24×7 cybersecurity operations center that provides real-time network monitoring, early cyber threat warnings and advisories, vulnerability identification and mitigation, and incident response for state, local, tribal, and territorial (SLTT) governments [20].

8.4 Mitigation from Denial of Service (DoS, DDoS, DRDoS) Attacks

A denial of service (DoS) attack overwhelms a website's servers by flooding them with bogus requests, making the websites unreachable or slowing the servers to a crawl. A distributed denial of service (DDoS) attack employs botnets—networks of computers infected with malware either with or without the knowledge of the computer owners. A distributed reflected denial of service (DRDoS) attack sends forged requests from a spoofed Internet Protocol (IP) address to a large number of computers that will reply to the requests [21].

In September 2012, Iran launched a major DDoS attack on American banks using botnets that involved thousands of compromised, high-powered application servers. The attackers used a new tool known as "itsoknoproblembro" (pronounced "it's OK, no problem, bro") to create peak floods exceeding 60 gigabits per second [22]. Bank of America, JPMorgan Chase, PNC Bank, U.S. Bank, and Wells Fargo Bank were among the financial institutions that suffered slowdowns and sporadic outages.

According to cybersecurity firm Prolexic Technologies, the total number of DDoS attacks increased 53 % in 2012 compared to a year earlier in 2011. Within a 3-month period from Q3 2012 to Q4 2012, the total number of infrastructure (Layer 3 and 4) attacks increased 17 %, the total number of application (Layer 7) attacks surged 72 %, the average attack duration rose 67 %, and the average bandwidth was up 20 %. Prolexic Technologies highlighted in its report that "Q4 2012 was defined by the increasing scale and diversity of DDoS attacks as well as the enduring nature of botnets" [23].

The most common network infrastructure (Layer 3 and 4) attack types are SYN floods and UDP floods, whereas the majority of flood traffic in the application (Layer 7) attack came in the form of GET floods and POST floods. Figure 8.2 shows the common DDoS attack vectors [23].

Various countermeasures against different types of DDoS attack exist in the Internet security community. Some common defenses are firewalls, routers, proxies, filtering, blackholing, and bandwidth over-provisioning [24–26]. However, as DDoS attacks are becoming more massive and sophisticated over the years, companies such as Prolexic Technologies, Tata Communications, and VeriSign have been offering their 24/7 monitoring and mitigation services. They provide a cloud-based DDoS mitigation service by "stopping a DDoS attack in the cloud before it reaches a customer's network" [27–29]. When a DDoS attack is detected, all traffic is routed through a "scrubbing center" which removes the bad traffic using filtering techniques and anti-DoS hardware devices, and establishes an Internet clean pipe to the customer's servers.

Nonetheless, no single countermeasure or mitigation service is 100 % efficacious. To stop DDoS attacks requires the entire international community to ACT—Achieve Cybersecurity Together. As the world is increasingly interconnected, everyone shares the responsibility of securing cyberspace.

The most effective solution to counter DDoS attacks is to drastically reduce the number of botnets and to eliminate their command-and-control (CnC) servers:

1. A powerful DDoS attack requires a large-scale botnet, preferably high-powered application servers or smartphones that are always on. By taking proactive measures in protecting our computer devices from malware and unauthorized access, we automatically reduce the scale of possible DDoS attacks. Proactive measures include applying system security patches promptly, installing firewalls

Infrastructure (Layer 3 and 4) Attacks	Application (Layer 7) Attacks
ACK Floods	HTTP GET Floods
DNS (Domain Name System) Attacks	HTTP POST Floods
FINPUSH Floods	NTP (Network Time Protocol) Flood
ICMP (Internet Control Message Protocol) Floods	PUSH Floods
IGMP (Internet Group Management Protocol) Floods	SSL GET Floods
RIP Flood	SSL POST Floods
SYN Floods	
SYN PUSH Flood	
TCP Fragment Floods	
TCP Reset Floods	
UDP Floods	
UDP Fragment Floods	

Fig. 8.2 Common DDoS attack vectors

and antivirus software, using strong passwords, and regularly checking system logs for suspicious activities.

2. A command-and-control (CnC) server is the brain of a botnet. In April 2011, the Federal Bureau of Investigation (FBI) dismantled the Coreflood botnet that had infected an estimated two million computers with malware. The FBI seized domain names, rerouted the botnet to FBI-controlled servers, and directed the zombie computers to stop the Coreflood malware [30]. In March 2012, Microsoft collaborated with the financial services industry and U.S. Marshals to seize CnC servers in Pennsylvania and Illinois, reducing the number of botnets controlling the Zeus malware that had infected more than 13 million computers worldwide [31]. Another success story—in the domain of spam attacks—is the takedown of the notorious Grum botnet in July 2012 [32]. Grum was the world's most prolific spam machine at its peak, generating 18 billion junk emails a day. Internet Service Providers (ISPs) in the Netherland, Panama, Russia, and Ukraine joined forces to knock down all of Grum's CnC servers in three days, and 50 % of worldwide spam vanished immediately [33].

Unless everyone—individual, business, and government—heeds the proactive countermeasures, DDoS is only going to get worse for everybody. In 2011, a Dirt Jumper v3 DDoS toolkit went on sale for as little as $150 on underground retail websites [34]. In 2012, a free-to-download High Orbit Ion Cannon (HOIC) program enables anyone online to launch a DDoS attack against a victim website [35].

8.5 Data Breach Prevention

At the 2012 RSA conference in San Francisco, Federal Bureau of Investigation (FBI) director Robert Mueller stated, "There are only two types of companies: those that have been hacked, and those that will be. Even that is merging into one category: those that have been hacked and will be again" [36]. Computer security executive Dmitri Alperovitch added, "I divide the entire set of Fortune Global 2000 firms into two categories: those that know they've been compromised and those that don't yet know" [37].

Cybercriminals have stolen intellectual property (IP) from high-profile companies such as Google and Adobe [38]. Retired FBI agent Shawn Henry spoke at the 2012 Black Hat Conference in Las Vegas: "We worked with one company that lost $1 billion worth of IP in the course of a couple of days—a decade of research. That is not an isolated event.... Your data is being held hostage, and the life of your organization is at risk" [39].

Some of the massive data breach incidents occurring between 2007 and early 2013 include:

1. In January 2007, up to 94 million Visa and MasterCard account numbers were stolen from TJX Companies dating back to 2005 and earlier, causing more than $68 million in fraud-related losses involving Visa cards alone [40].

2. In March 2011, computer and network security firm RSA suffered a massive data breach, jeopardizing the effectiveness of its SecurID system that is being used by more than 25,000 corporations and 40 million users around the world [41].
3. In April 2011, a cybercriminal stole the names, birth dates, and credit card numbers of 77 million customers on the Sony PlayStation Network [42].
4. In May 2011, cybercriminals illegally accessed 360,083 Citigroup customer accounts and withdrew $2.7 million from about 3,400 credit cards [43].
5. In January 2012, Global Payments suffered a data breach that put an estimated 50,000 Visa and MasterCard accounts at risk.
6. In June 2012, Russian cybercriminals stole 6.5 million LinkedIn passwords and posted them on an online forum [44]. Since the LinkedIn passwords were encoded using SHA-1—a cryptographic hash function with weak collision resistance, about half of the encrypted passwords have been decrypted and posted online.
7. In July 2012, Yahoo! Voices was hacked, resulting in the theft of 450,000 customer usernames and passwords [45].
8. In October 2012, Barnes & Noble disclosed that a PIN pad device used by customers to swipe credit and debit cards had been compromised at 63 of its national stores located in California, Connecticut, Florida, Illinois, Massachusetts, New Jersey, New York, Pennsylvania, and Rhode Island [46].
9. In January 2013, cybercriminals accessed Twitter user data and stole the usernames, email addresses, session tokens, and encrypted/salted versions of passwords for approximately 250,000 users [47].
10. In February 2013, personal information of more than 4,000 U.S. bank executives was stolen from the Federal Reserve System by exploiting a temporary vulnerability in a website vendor product [48].
11. In March 2013, cybercriminals gained access to cloud-storage service provider Evernote user information, including usernames, email addresses, and encrypted passwords. As a result, Evernote required all of its 50 million users to reset their passwords [49].

In 2012, the Verizon RISK (Response, Intelligence, Solutions, Knowledge) Team released a Data Breach Investigations Report in cooperation with the Australian Federal Police (AFP), Dutch National High Tech Crime Unit (NHTCU), Irish Reporting and Information Security Service (IRISS), U.K. Police Central e-Crime Unit (PCeU), and U.S. Secret Service (USSS) [50]. The most significant findings were:

1. The number of compromised data in 2011 skyrocketed to 174 million records, compared to 4 million a year before in 2010.
2. More than half (58 %) of the data breaches were tied to hacktivism (computer hacking + political activism).
3. 81 % of data breaches were due to hacking and 69 % of that involved malware.
4. Both hacking and malware incidents were up considerably by 31 and 20 % respectively.
5. An overwhelming 96 % of cyber attacks were unsophisticated and 97 % of breaches were avoidable through simple security measures such as firewalls and strong password protection.

6. 94 % of all data compromised involved servers.

In fact, the majority of data breach is preventable. In addition to using strong passwords, firewalls, and antivirus programs, there are three basic strategies that all businesses should follow:

1. One of the most common methods for breaching network security is SQL injection attack (SQLIA) as reported in the Yahoo! Voices data breach in July 2012 [51]. Structured Query Language (SQL) is the programming language used to manage data in a relational database management system (RDBMS). Cybercriminals inject unexpected or malformed SQL commands into the database in order to change its content or to dump its information to the attackers. The Open Web Application Security Project (OWASP) provides a clear, simple, and actionable guidance for preventing SQL Injection security flaws in application databases [52]. The primary defenses are using prepared statements (parameterized queries), using stored procedures, and escaping all user supplied input. Additional defenses are enforcing least privilege and performing white list input validation. Last but not least, strong encryption should be used to protect the sensitive information stored in the databases.

2. In the 2007 TJX Companies data breach incident, cybercriminals had intercepted wireless transfers of customer information at two Miami-area Marshalls stores since 2005 [40]. Global payment processing company First Data Corporation advices merchants to employ end-to-end encryption (E2EE) and tokenization: All merchants—whether they are brick-and-mortar, brick-and-click, or completely web-based—have both an obligation and an industry mandate to protect consumers' payment card data. The Payment Card Industry (PCI) Data Security Standards (DSS) provide guidelines on what merchants need to do to secure the sensitive data used in payment transactions. End-to-end encryption (E2EE) and tokenization solve for many of the vulnerabilities that exist in the payments processing chain. Encryption mitigates security weaknesses that exist when cardholder data has been captured but not yet authorized, and tokenization addresses security vulnerabilities after a transaction has been authorized. When combined, these two technologies provide an effective method for securing sensitive data wherever it exists throughout its lifecycle [53].

3. Data Loss Prevention (DLP) depends not only on cybsecurity technology but also on data management policies and human vigilance in implementing the policies. Shane MacDougall, champion of the social engineering "capture the flag" contest in DEF CON 20, made a phone call to a Wal-Mart store manager and successfully convinced him to divulge details of the store operations as well as computer security information. Wal-Mart spokesman Dan Fogleman told *CNNMoney*, "We emphasize techniques to avoid social engineering attacks in our training programs. We will be looking carefully at what took place and learn all we can from it in order to better protect our business" [54]. Perhaps one of the most heart-wrenching stories is about the security flaws in Apple and

Amazon's customer service that led to a devastating hack into *Wired* technology journalist Mat Honan in August 2012 [55]:

In the space of one hour, my entire digital life was destroyed. First my Google account was taken over, then deleted. Next my Twitter account was compromised, and used as a platform to broadcast racist and homophobic messages. And worst of all, my AppleID account was broken into, and my hackers used it to remotely erase all of the data on my iPhone, iPad, and MacBook.

Getting into Amazon let my hackers get into my Apple ID account, which helped them get into Gmail, which gave them access to Twitter.... But what happened to me exposes vital security flaws in several customer service systems, most notably Apple's and Amazon's. Apple tech support gave the hackers access to my iCloud account. Amazon tech support gave them the ability to see a piece of information—a partial credit card number—that Apple used to release information. In short, the very four digits that Amazon considers unimportant enough to display in the clear on the web are precisely the same ones that Apple considers secure enough to perform identity verification. The disconnect exposes flaws in data management policies endemic to the entire technology industry, and points to a looming nightmare as we enter the era of cloud computing and connected devices.

High-profile data breaches over the years ought to remind businesses that they are responsible for the safety of their customer's information. In November 2012, researchers Luigi Auriemma and Donato Ferrante found a serious vulnerability in the game "Call of Duty: Modern Warfare 3" as well as in the CryEngine 3 graphics platform. "Once you get access to the server, which is basically the interface with the company, you can get access to all of the information on the players through the server," Ferrante said. "In general, game companies don't seem to be very focused on security but rather on performance of the game itself" [56]. This attitude has to change. The Sony PlayStation Network data breach incident in April 2011 should have been a wake-up call for all game companies.

Facebook, with over 1 billion active monthly users, is an attractive target for cybercriminals. In January 2013, a sophisticated cyber attack targeted Facebook when its employees visited a mobile developer website that was compromised [57]. The Facebook Security team flagged a suspicious domain in their corporate DNS logs and discovered the presence of malware on several company laptops. However, they found no evidence that any Facebook user data was compromised.

8.6 Fighting Back Against Phishing and Spoofing

Phishing is an example of social engineering designed to manipulate people into divulging confidential information. The most common form of phishing is an email that appears to have come from a legitimate organization or known individual. Similar to Caller ID spoofing, the sender's email address is often forged in order to hide the real identity and origin of the sender.

The recipient of a phishing email is encouraged to open an attachment that would install malware on the victim's computer, or to click on a link that directs the user to a fake website whose look and feel are almost identical to the legitimate site.

In February 2011, *Contagio* published the details of a targeted phishing attack (aka spear phishing) against personal Gmail accounts of U.S. military and government employees and associates [58]. The spoofed emails appeared to have come from .gov and .mil domains, and they contained a link to a fake Gmail timed-out re-login page for the attackers to harvest the victim's credentials.

A more sophisticated hack involves DNS spoofing, DNS cache poisoning, or DNS hijacking whereby the Domain Name System (DNS) used to translate domain names into IP addresses is compromised. In 2009, Brazil's largest bank had its domain name redirected to a criminal's computer server for four hours, fooling customers who tried to log into their accounts [59]. Between 2008 and 2012, about 4 million computers worldwide were infected with the DNSChanger malware that redirected the victims to spoofed websites [60]. In November 2012, China intentionally DNS poisoned www.google.com and all other Google subdomains for political reasons [61].

The Anti-Phishing Working Group (APWG) released a phishing activity trends report in September 2012 [62]. The report reveals some staggering numbers:

1. The number of unique phishing sites reached an all-time monthly high of 63,253 in April 2012, with an average of 58,409 in Q2 2012.
2. The total number of URLs used to host phishing attacks increased 7 % to 175,229 in Q2 2012, up from 164,023 in Q1 2012.
3. The U.S. remained the top hosting country of phishing-based Trojan horse virus in Q2 2012: United States 46 % versus 2nd place Russia 12 % in April, United States 78 % versus 2nd place United Kingdom 4 % in May, and United States 55 % versus 2nd place France 11 % in June.

To fight back against phishing and spoofing, businesses, governments, and individuals must take all possible proactive measures:

1. Sender Policy Framework (SPF) and DomainKeys Identified Mail (DKIM) have helped reduce the ease of spoofing. SPF allows the domain owners to specify which computers are authorized to send emails with sender addresses in that domain, whereas DKIM enables an email to carry a digital signature specifying its genuine domain name. Email service providers such as Gmail and Yahoo! have implemented DKIM since 2008. In collaboration with eBay and PayPal, Gmail automatically discards all incoming emails that claim to be coming from ebay.com or paypal.com if they cannot be verified successfully with DKIM [63]. However, mathematician Zachary Harris discovered in 2012 that weak cryptographic keys used in DKIM exposed a massive security hole in Apple, Amazon.com, eBay, Gmail, HSBC, LinkedIn, Microsoft, PayPal, Twitter, US Bank, Yahoo!, and some other large organizations [64]. U.S. Department of Homeland Security's United States Computer Emergency Readiness Team (US-CERT) issued a vulnerability note in October 2012 warning that "DomainKeys Identified Mail (DKIM) Verifiers may inappropriately convey message trust when messages are signed using keys that are too weak (<1024 bits) or that are marked as test keys" [65]. Businesses should implement stronger cybersecurity with 1024-bit or 2048-bit cryptographic keys.

2. US-CERT has been collecting phishing email messages and website locations so that the information can help people avoid becoming victims of phishing scams [66]. Recipients of phishing emails are encouraged to forward them to phishing-report@us-cert.gov. In addition, social network users should forward all suspected phishing messages to phish@fb.com for Facebook, spoof@ebay.com for eBay, and @spam for Twitter. We should also report on fraudulent web pages that are designed to look like the legitimate websites in attempt to steal users' personal information. Reports can be submitted to the Google Safe Browsing Team [67]: http://www.google.com/safebrowsing/report_phish/.

3. Commercial websites can discourage phishing by making it easier for users to distinguish between genuine and fake sites by authenticating the site to the user. The method is known as two-way authentication, mutual authentication, or bidirectional authentications. For example, Capital One 360 (formerly ING Direct) employs two-factor, two-way authentication [68]. First, a customer signs in with an username and password. Second, ING Direct displays a secret phrase and picture that only the customer knows about when setting up the account. Third, the customer enters a six-digit passcode to gain access to the account. If a commercial website registers the IP address of the customer's computer, technically it can authenticate itself to the user even before asking for their username and password. Two-way mutual authentication is an effective guard against DNS spoofing.

8.7 Password Protection and Security Questions

In September 2008, 20-year-old college student David Kernell hacked into Republican vice presidential candidate Sarah Palin's Yahoo! email account to look for information that would derail her campaign [69]. Kernell managed to reset Palin's account password by entering her birth date and correctly answering the security question "Where did you meet your spouse?" It only took Kernell 45 minutes on Wikipedia and Google search to find the correct answer.

In July 2012, CNet analyzed the most frequently used passwords that surfaced in the Yahoo! Voices data breach. Among more than 450,000 stolen login credentials, the most common passwords in descending order of popularity were 123456, password, 111111, welcome, ninja, freedom, f*ck, baseball, superman, 000000, America, winner, starwars, batman, spiderman, lakers, maverick, ncc1701, startrek, and ncc1701a [70].

These and many other cybercriminal stories highlight the vulnerability of weak password protection and inadequacy of security questions. The best practices for protecting our login credentials are:

1. Turn on two-factor authentication on Google, Facebook, and other websites. Many banking sites such as ING Direct require two-factor authentication automatically.

2. When coming up with a password, avoid dictionary words, acronyms, and abbreviations. A password should be difficult to guess but easy to remember

so that it does not need to be written down anywhere. For example, if Jack and Jill are a married couple who visit Disneyland with their three children twice every year, a strong and easy-to-remember password would be something like "J&J3di2yr" which includes mixed case letters, numbers, and punctuations. Changing your password regularly is also made simple by slightly rearranging the password phrase. For instance, "J&J3di2yr" could become "JJ&3di2yr".

3. Do not use the same password across multiple websites. Instead, create an unique password for every site. It may sound like a daunting task if you have an account on many different websites, but there is a simple two-step solution. Step one: Create a strong password stem. Step two: Append a site-specific phrase to the stem. The sample password above, "J&J3di2yr", is an example of a strong password stem. Now, if Jack and Jill share a Gmail account that Jill created on the day her mother-in-law was visiting from England, a strong and easy-to-remember password for their Gmail account would be something like "J&J3di2yrMotherLand". Similarly, one of Jack and Jill's bank accounts might have the password "J&J3di2yrBettyBoop" because they ate at a local "Betty Boop" restaurant on the day they opened that particular bank account. The basic idea is to combine a strong password stem with phrases from associative memory or good imagination.

4. Do not answer the online security questions straightforwardly. Instead, treat the answers as passwords, give a long answer, or simply be creative. The following are some excellent examples:

   (a) Where did you meet your spouse? Answer: J&Jvt1980lunch
   (Jack and Jill met at Virginia Tech in 1980 during lunch).
   (b) What was the name of your first school? Answer: Ismellcheese
   (Your first school reminds you of the smell of cheese, for whatever reason).
   (c) What is your pet's name? Answer: Squarerootofminus178
   (The square root of a negative number is an imaginary number, meaning I don't have a pet or I have an imaginary pet).

8.8 Software Upgrades and Security Patches

Upgrading the computer operating system to the latest version is highly recommended for security reasons. A 64-bit computer running Windows 7 and Internet Explorer 9, for example, is inherently more secure than an older version of Windows operating system because of new security technologies such as Address Space Layout Randomization (ASLR), Data Execution Prevention (DEP), and SmartScreen Filter [71].

Installing application software patches is as equally important as performing regular operating system updates. A vulnerability in earlier versions of Adobe Reader and Acrobat allowed cybercriminals to inject malicious code in PDF documents, and the hack became so prevalent that it represented 80 % of all exploits in 2009 [72].

In April 2012, cybercriminals exploited a Java vulnerability and infected more than half a million Apple computers by downloading malware without prompting [73].

With 3 billion devices running Java, the hack is alarming. In most cases, installing security patches is sufficient. In January 2013, however, US-CERT urged users to disable Java on their Web browsers even after Oracle fixed a serious security flaw in its Java software [74].

Upgrading software and applying security patches should be done at home or in the office, and never at hotels using public Wi-Fi. A virus such as "Flashback" may look like a normal Adobe Flash browser plug-in, and a malware program may be named inconspicuously like "Android System Update 4.1.2.apk". In May 2012, the FBI issued the following warning regarding hotel Wi-Fi use [75]:

> Recently, there have been instances of travelers' laptops being infected with malicious software while using hotel Internet connections. In these instances, the traveler was attempting to setup the hotel room Internet connection and was presented with a pop-up window notifying the user to update a widely-used software product. If the user clicked to accept and install the update, malicious software was installed on the laptop. The pop-up window appeared to be offering a routine update to a legitimate software product for which updates are frequently available.
>
> The FBI recommends that all government, private industry, and academic personnel who travel abroad take extra caution before updating software products on their hotel Internet connection. Checking the author or digital certificate of any prompted update to see if it corresponds to the software vendor may reveal an attempted attack. The FBI also recommends that travelers perform software updates on laptops immediately before traveling, and that they download software updates directly from the software vendor's Web site if updates are necessary while abroad.

The sophisticated Stuxnet worm has demonstrated that signed digital certificates can be stolen to make a malware app look legitimate [76]. The safest software updates come directly from the software vendor's website, unless the site is hijacked by DNS spoofing.

8.9 Fake Software and Free Downloads

In September 2011, Microsoft issued a severe security alert of a fake security protection software program that displays alerts for non-existent threats on a computer in order to entice the user to download the malware [77] (see Fig. 8.3).

Shortly after the popular game "Angry Birds Space" was released in March 2012, fake apps containing the Trojan horse virus began to surface in Android markets. One counterfeit app appears to be a fully functional Angry Birds game, but it installs a virus on the user's smartphone or tablet [78]. Another fake app disguises itself as Angry Birds and turns the infected smartphone into a SpamSoldier bot for its spam campaign [79].

In August 2012, *Network World* reported a recent study claiming that over 90 % of the top 100 paid apps in the Apple App Store and Google Play Android Market have been pirated, hacked, malware-laden, and then given away for free [80]. The old sayings "You get what you pay for" and "If it's too good to be true, it probably is" are certainly appropriate here. Extra precaution is warranted before downloading any "free" games, music, movies, or software.

Fig. 8.3   A fake scanner interface of the "security protection" malware

The same vigilance applies to businesses. To attract customers, some retailers preinstall on new computers free software and games from questionable sources and distribution channels. Microsoft Security Intelligence Report recently revealed a disturbing finding that "malware has been discovered preinstalled on computers sold at retail" [81].

In September 2012, Microsoft disrupted the spread of the Nitol botnet malware embedded in counterfeit Windows operating system sold with some new computers [82]. "Consumers should exercise their right to demand that resellers provide them with non-counterfeit products free of malware," said Richard Domingues Boscovich, Assistant General Counsel at the Microsoft Digital Crimes Unit.

To ensure that a computer running Windows is free of malware, Microsoft offers a free security program—Microsoft Safety Scanner—that removes viruses, spyware, and other malicious software [83]. The safety scanner is constantly updated with the latest anti-malware definitions, therefore it expires 10 days after each download, and it has to be re-downloaded and re-run periodically.

8.10   Smartphone Security Protection

In 1957, a blind seven-year old boy named Joe Engressia (aka Joybubbles) with an IQ of 172 used his unusual auditory gifts to hack the analog POTS (Plain Old Telephone Service) and became a nerve center of the "phone phreaks" subculture in 1970s [84]. Joybubbles could dial by using the hookswitch like a telegraph key. He could place free long distance phone calls by whistling the proper tones at

2,600 Hz into any telephone. The world of communication has since changed to digital, and phreaks were precursors of today's computer hackers.

In February 2012, technology and market research company Forrester Research estimated that one billion people will own smartphones by 2016 [85]. These mobile phones are powerful little computers that are always on, 24/7. Consequently, they are the perfect targets for cybercriminals to steal information and to turn the phones into a botnet for launching a DDoS attack or spam campaign.

Cyber attacks on mobile phones rose by a whopping 500 % in 2012, according to McAfee [86]; and 10 % of all adults surveyed have experienced cybercrime on their mobile devices, according to Symantec [87]. The following are some of the exploits that have been discovered so far:

1. March 2005: The Commwarrior. A virus replicates itself by sending multimedia messages (MMS) to people on the phone's contacts list [88].
2. April 2012: A counterfeit game based on "Angry Birds Space" installs a Trojan horse virus on victims' smartphones [78].
3. July 2012: A security flaw in the Android framework in Version 4.0.4 and below could be exploited by a rootkit, and no existing mobile security software was able to detect it [89].
4. July 2012: Vulnerabilities in the "near field communications (NFC)" features on some smartphones allow a tag with an embedded NFC chip to push a webpage to victims' phones, exploit a browser bug, and obtain unauthorized access [90].
5. December 2012: An exploit targeted at smartphones that use certain Exynos processors can let Android malware apps steal and delete all the data on victims' phones [91].
6. December 2012: Two counterfeit Android games, based on "Angry Birds Star Wars" and "The Need for Speed Most Wanted," were infected with malware that can turn victims' smartphones into a botnet for launching a mobile SMS spam campaign [92].

"Your cell phone is communicating completely digital; it's part of the Internet," said Army Gen. Keith Alexander, director of the NSA and commander of the U.S. Cyber Command. "The attack surfaces for adversaries to get on the Internet now include all those mobile devices.... The mobile security situation lags. It's far behind" [93].

The current design of mobile devices makes differentiating legitimate sites from malicious ones a tricky task. "No matter how tantalizing a link might look on a desktop, there are cues that you shouldn't go there, such as an address that just doesn't look safe," said Hugh Thompson, Blue Coat's senior vice president and chief security strategist. "When you click a link on a mobile phone, it's harder to know what form of Russian roulette they're playing" [94].

In June 2012, the Defense Advanced Research Projects Agency (DARPA) assigned cybersecurity firm Invincea a $21 million research grant to fortify Android-based phones and tablets for use by the military personnel [95]. New research ideas include creating a virtual run-time environment separating military applications from other commercial software such as Facebook, Twitter, Skype, and games running on the smartphones.

For civilians and military personnel alike, the Federal Communications Commission (FCC) has released a "Smartphone Security Checker" to help consumers secure their mobile devices [96]. The general security checklist is as follows:

1. Set PINs and passwords to prevent unauthorized access to your phone. Configure your phone to automatically lock after 5 min or less when your phone is idle, as well as use the SIM password capability available on most smartphones.
2. Do not modify your smartphone's security settings. Tampering with your phone's factory settings, jailbreaking, or rooting your phone undermines the built-in security features offered by your wireless service and smartphone, while making it more susceptible to an attack.
3. Backup and secure all of the data stored on your phone.
4. Only install mobile apps from trusted sources.
5. Understand app permissions before accepting them. Be cautious about granting applications access to personal information on your phone or otherwise letting the application have access to perform functions on your phone.
6. Install anti-theft security protection apps that enable remote location and wiping. Some carriers offer a free "remote wipe" service that allows users to delete all of the data from a lost or stolen device to prevent data or identity theft. You cannot rely solely on passcode to protect your smartphone's content. It was discovered in January 2013 that a security flaw in Apple's iOS 6.1 allows anyone to bypass your iPhone password lock [97].
7. Keep your phone's software up-to-date by enabling automatic updates or accepting updates when prompted from your service provider, operating system provider, device manufacturer, or application provider.
8. Be smart on open Wi-Fi networks. When you access a public Wi-Fi network, your phone can be an easy target of cybercriminals. Always be aware when clicking on web links and be particularly cautious if you are asked to enter account or login information.
9. Wipe data on your old phone before you donate, resell, or recycle it. To protect your privacy, completely erase data off of your phone and reset the phone to its initial factory settings.
10. Report a stolen smartphone. The major wireless service providers, in coordination with the FCC, have established a stolen phone database. This will provide notice to all the major wireless service providers that the phone has been stolen and will allow for remote "bricking" of the phone so that it cannot be activated on any wireless network without your permission.

As if taking a page from a Hollywood script for *Mission Impossible* or a James Bond movie, GPS (Global Positioning System) spoofing and IMSI (International Mobile Subscriber Identity) catchers are security concerns for the military and law enforcement:

1. GPS spoofing broadcasts a set of normal GPS signals but different from the GPS satellites in order to deceive a GPS receiver, be it a smartphone or a

wireless drone. In December 2011, Iranian electronic warfare engineers claimed to have intentionally broadcasted misguided GPS signals to hijack a CIA drone (US RQ-170 Sentinel) and guide it to an intact landing inside Iran [98].
2. An IMSI catcher (aka Stingray) is a virtual base transceiver station (VBTS) that can identify the IMSI of nearby GSM mobile phones and intercept their calls. Since the GSM (Global System for Mobile Communications) specification requires the handset to authenticate to the network but vice versa, an IMSI catcher can easily trick mobile phones into thinking that they are connected to a legitimate service provider's cellular network [99].

Similar to tackling DNS spoofing, an effective countermeasure to GPS spoofing and IMSI catchers is two-way mutual authentication between the smartphones and the cell towers or base transceiver stations (BTS).

Smartphone security protection and hacking will likely continue to be a cat-and-mouse game. "It's possible for something to go wrong on the scale of a big wireless network because of a coding mistake in an operating system or an application, and it's very hard to diagnose and fix," said David Fritz, senior technical staff at Sandia National Laboratories. "You can't possibly read through 15 million lines of code and understand every possible interaction between all these devices and the network" [100].

8.11 Cybersecurity Awareness: Everyone's Responsibility

Wired technology journalist Mat Honan wrote that besides Amazon and Apple's faults, he should have used two-factor authentication for his Google account and regularly backed up the data on his MacBook [55]. His story reminds both businesses and consumers that cybersecurity is everyone's responsibility.

We cannot rely solely on antivirus software. During a four-month long cyber attack on *The New York Times*, cybercriminals installed 45 pieces of custom malware on the network between October 2012 and January 2013. Symantec antivirus products installed at the Times were able to identify and quarantine only one piece of malware, missing all 44 others [101]. Symantec issued a follow-up statement, "Turning on only the signature-based anti-virus components of endpoint solutions alone are not enough in a world that is changing daily from attacks and threats. We encourage customers to be very aggressive in deploying solutions that offer a combined approach to security. Anti-virus software alone is not enough" [102].

"Even the most modern version of antivirus software doesn't give consumers or enterprises what they need to compete in the hacker world," said Dave Aitel, CEO of software security firm Immunity. "Deep down, nothing is as good has having a proper awareness about what's going on in your network" [103]. In February 2013, NBC.com and related sites were hacked and infected with malware that redirected visitors to malicious websites. "This morning, NBC.com was hacked and embedded with malicious iframe code that spread the Citadel Trojan. It was detected as Backdoor.Agent.RS. ... The NBC web site was compromised for about 15 min and the actual iframe with the malicious redirect was embedded in a javascript file located on the NBC.com web server," said an NBC spokesperson [104].

More than 400 million people trust Google with their emails, and 50 million people store files in the cloud using the Dropbox service [105]. In February 2013, the U.S. Secret Service investigated the hacking and publication of private photographs and emails between members of the Bush family, including former Presidents George H.W. Bush and George W. Bush [106].

Cyber espionage and cyber criminal activities have forced businesses to take extraordinary measures to safeguard customers' information and prevent data breach. Google, for instance, hired DARPA director Regina Dugan in 2012 to fill a senior executive position. "Regina is a technical pioneer who brought the future of technology to the military during her time at DARPA," said a Google spokeswoman in an email to *Computerworld*. "She will be a real asset to Google, and we are thrilled she is joining the team" [107].

Dugan spoke at the TED 2012 conference in Long Beach, California about the spirit of scientists and engineers at DARPA: "When you remove the fear of failure, impossible things suddenly become possible" [108]. Among many of the DARPA inventions, she also talked about lightning that occurs in nature: "There are 44 lightning strikes per second around the globe. Each lightning bolt heats the air to 44,000 degrees Fahrenheit, hotter than the surface of the sun. What if we could use these electromagnetic pulses as beacons—beacons in a moving network of powerful transmitters. Experiments suggest that lightning could be the next GPS" [108].

Perhaps in the future, lightning will be used as a countermeasure to GPS spoofing. But in the meantime, companies like Facebook and Google have been enlisting hackers to find security holes in their products. In August 2011, Facebook paid out more than $40,000 within a month under its new "bug bounty" security initiative [109]. The company offers a minimum reward of $500 [110] and publicly thanks the "white-hat hackers" on the Facebook page [111]. In October 2012, Google awarded the top $60,000 prize to teenage hacker "Pinkie Pie" for uncovering a vulnerability in the Chrome browser [112]. It was the teen's second win. 10 hours after the bug was exposed, Google issued the Chrome security fixes and announced on its official blog: "Congratulations to Pinkie Pie, returning to the fray with another beautiful piece of work!" [113].

In the spirit of President John F. Kennedy, one may proclaim: "Ask not what cybersecurity can do for you, ask what you can do for cybersecurity".

References

1. **Schiff, et al.** H.R.2290 – Cyber-Security Enhancement Act of 2007 (Introduced in House—IH). [Online] The Library of Congress, May 14, 2007. http://thomas.loc.gov/cgi-bin/query/z?c110:H.R.2290:.
2. **Ackerman, et al.** H.R.7007—National Commission on American Cybersecurity Act of 2008. [Online] The Library of Congress, September 23, 2008. http://thomas.loc.gov/cgi-bin/query/z?c110:H.R.7007:.

3. **Rockefeller, et al.** S.773—Cybersecurity Act of 2009. [Online] The Library of Congress, April 1, 2009. http://thomas.loc.gov/cgi-bin/query/z?c111:S.773:.

4. **Reid, et al.** S.21—Cyber Security and American Cyber Competitiveness Act of 2011. [Online] The Library of Congress, January 25, 2011. http://thomas.loc.gov/cgi-bin/query/z?c112:S.21:.

5. **Lieberman, et al.** S.2105 – Cybersecurity Act of 2012. [Online] The Library of Congress, February 14, 2012. http://thomas.loc.gov/cgi-bin/query/z?c112:S.2105:.

6. **Lieberman, Joe, Collins, Susan and Carper, Tom.** A gold standard in cyber-defense. [Online] The Washington Post, July 7, 2011. http://www.washingtonpost.com/opinions/a-gold-standard-in-cyber-defense/2011/07/01/gIQAjsZk2H_story.html.

7. **Nagesh, Gautham.** Sen. Rockefeller presses Congress to pass cybersecurity legislation. [Online] The Hill, January 31, 2012. http://thehill.com/blogs/hillicon-valley/technology/207729-rockefeller-presses-congress-to-pass-cybersecurity-legislation.

8. **Kain, Erik.** Does The Cybersecurity Act Of 2012 Mark The Beginning Of The War On Cyber-terrorism? [Online] Forbes, February 22, 2012. http://www.forbes.com/sites/erikkain/2012/02/22/does-the-cybersecurity-act-of-2012-mark-the-beginning-of-the-war-on-cyber-terrorism/.

9. **Stiennon, Richard.** Rockefeller's Cybersecurity Act of 2010: A Very Bad Bill. [Online] Forbes, May 4, 2010. http://www.forbes.com/sites/firewall/2010/05/04/rockefellers-cybersecurity-act-of-2010-a-very-bad-bill/.

10. **Brito, Jerry and Watkins, Tate.** Wired Opinion: Cyberwar Is the New Yellowcake. [Online] Wired, February 14, 2012. http://www.wired.com/threatlevel/2012/02/yellowcake-and-cyberwar/.

11. **Slngel, Ryan.** NSA Must Examine All Internet Traffic to Prevent Cyber Nine-Eleven, Top Spy Says. [Online] Wired, January 15, 2008. http://www.wired.com/threatlevel/2008/01/feds-must-exami/.

12. **Obama, Barack.** Executive Order—Improving Critical Infrastructure Cybersecurity. [Online] The White House, February 12, 2013. http://www.whitehouse.gov/the-press-office/2013/02/12/executive-order-improving-critical-infrastructure-cybersecurity.

13. Cyberspace Policy Review: Assuring a Trusted and Resilient Information and Communications Infrastructure. [Online] The White House, May 8, 2009. http://www.whitehouse.gov/assets/documents/Cyberspace_Policy_Review_final.pdf.

14. **National Security Council.** The Comprehensive National Cybersecurity Initiative. [Online] The White House. [Cited: January 18, 2013.] http://www.whitehouse.gov/cybersecurity/comprehensive-national-cybersecurity-initiative.

15. **The White House.** National Initiative for Cybersecurity Education (NICE) Relationship to President's Education Agenda. [Online] The White House, April 19, 2010. http://www.whitehouse.gov/sites/default/files/rss_viewer/cybersecurity_niceeducation.pdf.

16. **Information Assurance Support Environment (IASE).** Cyber Awareness Challenge. [Online] U.S. Department of Defense. [Cited: January 21, 2013.] http://iase.disa.mil/eta/cyberchallenge/launchPage.htm.

17. **Homeland Security.** National Cyber Security Awareness Month. [Online] U.S. Department of Homeland Security. [Cited: January 18, 2013.] http://www.dhs.gov/national-cyber-security-awareness-month.

18. **Homeland Security.** Cyber Storm: Securing Cyber Space. [Online] U.S. Department of Homeland Security. [Cited: January 18, 2013.] http://www.dhs.gov/cyber-storm-securing-cyber-space.

19. **National Cybersecurity Alliance (NCSA).** StaySafeOnline.org. [Online] National Cybersecurity Alliance (NCSA). [Cited: January 18, 2013.] http://www.staysafeonline.org/.

20. **Center for Internet Security.** Multi-State Information Sharing and Analysis Center (MS-ISAC). [Online] Center for Internet Security. [Cited: January 18, 2013.] http://msisac.cisecurity.org/.

21. **Patrikakis, Charalampos, Masikos, Michalis and Zouraraki, Olga.** Distributed Denial of Service Attacks. [Online] The Internet Protocol Journal, December 2004. http://www.cisco.com/web/about/ac123/ac147/archived_issues/ipj_7-4/dos_attacks.html.

22. **Goodin, Dan.** DDoS attacks on major US banks are no Stuxnet—here's why. [Online] ArsTechnica, October 3, 2012. http://arstechnica.com/security/2012/10/ddos-attacks-against-major-us-banks-no-stuxnet/.

23. **Prolexic.** Prolexic Quarterly Global DDoS Attack Report. [Online] Prolexic, Q4 2012. http://www.prolexic.com/knowledge-center-ddos-attack-report-2012-q4/pr.html.
24. **Eddy, W.** TCP SYN Flooding Attacks and Common Mitigations. [Online] The Internet Engineering Task Force (IETF), August 2007. http://tools.ietf.org/html/rfc4987.
25. **VeriSign.** DDoS Mitigation - Best Practices for a Rapidly Changing Threat Landscape Whitepaper. [Online] VeriSign, 2012. http://www.verisigninc.com/en_US/products-and-services/network-intelligence-availability/nia-information-center/ddos-best-practice-confirmation/index.xhtml.
26. **Cisco.** Defeating DDOS Attacks. [Online] Cisco. [Cited: January 20, 2013.] http://www.cisco.com/en/US/prod/collateral/vpndevc/ps5879/ps6264/ps5888/prod_white_paper0900aecd8011e927.html.
27. **VeriSign.** VeriSign Internet Defense Network Enhanced With New DDoS Monitoring Service. [Online] Reuters, September 10, 2009. http://www.reuters.com/article/2009/09/10/idUS126052+10-Sep-2009+MW20090910.
28. **Prolexic.** Prolexic Issues Mitigation, Detection Rules for Critical DDoS Threat Used in Banking Attacks. [Online] PresseBox, January 3, 2013. http://www.pressebox.com/inactive/prolexic-technologies/Prolexic-Issues-Mitigation-Detection-Rules-for-Critical-DDoS-Threat-Used-in-Banking-Attacks/boxid/564817.
29. **Tata Communications.** Cloud-based security services. [Online] Tata Communications. [Cited: January 20, 2013.] http://security.tatacommunications.com/cloud.asp.
30. **Mueller, Robert S. III.** Robert S. Mueller, III Speech at RSA Cyber Security Conference. [Online] Federal Bureau of Investigation, March 1, 2012. http://www.fbi.gov/news/speeches/combating-threats-in-the-cyber-world-outsmarting-terrorists-hackers-and-spies.
31. **Microsoft.** Microsoft Joins Financial Services Industry to Disrupt Massive Zeus Cybercrime Operation That Fuels Worldwide Fraud and Identity Theft. [Online] Microsoft, March 25, 2012. http://www.microsoft.com/en-us/news/press/2012/mar12/03-25CybercrimePR.aspx.
32. **Mushtaq, Atif.** Grum, World's Third-Largest Botnet, Knocked Down. [Online] FireEye, July 18, 2012. http://blog.fireeye.com/research/2012/07/grum-botnet-no-longer-safe-havens.html.
33. **Cowley, Stacy.** Grum takedown: '50% of worldwide spam is gone'. [Online] CNNMoney, July 19, 2012. http://money.cnn.com/2012/07/19/technology/grum-spam-botnet/index.htm.
34. **Prolexic.** Prolexic Issues Dirt Jumper Threat Advisory and Releases Free Security Scanner. [Online] PRWeb, December 29, 2011. http://www.prweb.com/releases/2011/12/prweb9067808.htm.
35. **Breeden, John II.** Hackers' new super weapon adds firepower to DDOS. [Online] GCN, October 24, 2012. http://gcn.com/Articles/2012/10/24/Hackers-new-super-weapon-adds-firepower-to-DDOS.aspx.
36. **Cowley, Stacy.** FBI Director: Cybercrime will eclipse terrorism. [Online] CNNMoney, March 2, 2012. http://money.cnn.com/2012/03/02/technology/fbi_cybersecurity/index.htm.
37. **Perlroth, Nicole.** Some Victims of Online Hacking Edge Into the Light. [Online] The New York Times, February 20, 2013. http://www.nytimes.com/2013/02/21/technology/hacking-victims-edge-into-light.html
38. **Zetter, Kim.** Google Hack Attack Was Ultra Sophisticated, New Details Show. [Online] Wired, January 14, 2010. http://www.wired.com/threatlevel/2010/01/operation-aurora/.
39. **Cowley, Stacy.** Former FBI cyber cop worries about a digital 9/11. [Online] CNN, July 25, 2012. http://money.cnn.com/2012/07/25/technology/blackhat-shawn-henry/index.htm.
40. **Jewell, Mark.** TJX breach could top 94 million accounts. [Online] NBC News, October 24, 2007. http://www.msnbc.msn.com/id/21454847/ns/technology_and_science-security/t/tjx-breach-could-top-million-accounts/.
41. **Hickins, Michael and Clark, Don.** Questions Over Break-In at Security Firm RSA. [Online] The Wall Street Journal, March 18, 2011. http://online.wsj.com/article/SB10001424052748703512404576208983743029392.html.
42. **Wingfield, Nick, Sherr, Ian and Worthen, Ben.** Hacker Raids Sony Videogame Network. [Online] The Wall Street Journal, April 27, 2011. http://online.wsj.com/article/SB10001424052748703377810457628736250377653 4.html.

43. **Smith, Aaron.** Citi: Millions stolen in May hack attack. [Online] CNNMoney, June 27, 2011. http://money.cnn.com/2011/06/27/technology/citi_credit_card/index.htm.
44. **Goldman, David.** More than 6 million LinkedIn passwords stolen. [Online] CNNMoney, June 7, 2012. http://money.cnn.com/2012/06/06/technology/linkedin-password-hack/index.htm.
45. **Gross, Doug.** Yahoo hacked, 450,000 passwords posted online. [Online] CNN, July 13, 2012. http://www.cnn.com/2012/07/12/tech/web/yahoo-users-hacked/index.html?hpt=hp_t1.
46. **Riley, Charles.** Barnes & Noble customer data stolen. [Online] CNNMoney, October 24, 2012. http://money.cnn.com/2012/10/24/technology/barnes–noble-hack/index.html.
47. **Lord, Bob.** Keeping our users secure. [Online] Twitter Blog, February 1, 2013. http://blog.twitter.com/2013/02/keeping-our-users-secure.html.
48. **Bull, Alister and Finkle, Jim.** Fed says internal site breached by hackers, no critical functions affected. [Online] Reuters, February 6, 2013. http://www.reuters.com/article/2013/02/06/net-us-usa-fed-hackers-idUSBRE91501920130206.
49. **Engberg, Dave.** Security Notice: Service-wide Password Reset. [Online] The Evernote Blog, March 2, 2013. http://blog.evernote.com/blog/2013/03/02/security-notice-service-wide-password-reset/
50. **Verizon RISK Team.** 2012 Data Breach Investigations Report. [Online] Verizon, 2012. http://www.verizonbusiness.com/resources/reports/rp_data-breach-investigations-report-2012_en_xg.pdf.
51. **Menegaz, Gery.** SQL Injection Attack: What is it, and how to prevent it. [Online] ZDNet, July 13, 2012. http://www.zdnet.com/sql-injection-attack-what-is-it-and-how-to-prevent-it-7000000881/.
52. **OWASP.** SQL Injection Prevention Cheat Sheet. [Online] The Open Web Application Security Project, December 6, 2012. https://www.owasp.org/index.php/SQL_Injection_Prevention_Cheat_Sheet.
53. **First Data Corporation.** What Data Thieves Don't Want You to Know: The Facts About Encryption. [Online] First Data Corporation, 2012. http://www.firstdata.com/downloads/thought-leadership/TokenizationEncryptionWP.pdf.
54. **Cowley, Stacy.** How a lying 'social engineer' hacked Wal-Mart. [Online] CNNMoney, August 8, 2012. http://money.cnn.com/2012/08/07/technology/walmart-hack-defcon/index.htm.
55. **Honan, Mat.** How Apple and Amazon Security Flaws Led to My Epic Hacking. [Online] Wired, August 6, 2012. http://www.wired.com/gadgetlab/2012/08/apple-amazon-mat-honan-hacking/all/.
56. **Kirk, Jeremy.** Researchers find vulnerability in Call of Duty: Modern Warfare 3. [Online] CSO, November 9, 2012. http://www.csoonline.com/article/721133/researchers-find-vulnerability-in-call-of-duty-modern-warfare-3.
57. **Facebook Security.** Protecting People On Facebook. [Online] Facebook, February 15, 2013. https://www.facebook.com/notes/facebook-security/protecting-people-on-facebook/10151249208250766.
58. **Mila.** Targeted attacks against personal accounts of military, government employees and associates. [Online] Contagio, February 17, 2011. http://contagiodump.blogspot.com/2011/02/targeted-attacks-against-personal.html.
59. **Kelly, Suzanne and Benson, Pam.** U.S. gears up for cyberwar amid conflicting ideas on how to fight it. [Online] CNN, February 24, 2012. http://security.blogs.cnn.com/2012/02/24/u-s-gears-up-for-cyberwar-amid-conflicting-ideas-on-how-to-fight-it/.
60. **Kim, Erin.** Internet blackout for thousands begins Monday. [Online] CNNMoney, July 9, 2012. http://money.cnn.com/2012/07/06/technology/dnschanger/index.htm.
61. **Whittaker, Zack.** Google services 'disrupted' in China; traffic declines rapidly. [Online] ZDNet, November 9, 2012. http://www.zdnet.com/google-services-disrupted-in-china-traffic-declines-rapidly-7000007195/.
62. **APWG.** Phishing Activity Trends Report (2nd Quarter 2012). *Anti-Phishing Working Group (APWG).* [Online] September 2012. http://docs.apwg.org/reports/apwg_trends_report_q2_2012.pdf.
63. **Taylor, Brad.** Fighting phishing with eBay and PayPal. [Online] Official Gmail Blog, July 8, 2008. http://gmailblog.blogspot.com/2008/07/fighting-phishing-with-ebay-and-paypal.html#!/2008/07/fighting-phishing-with-ebay-and-paypal.html.

64. **Zetter, Kim.** How a Google Headhunter's E-Mail Unraveled a Massive Net Security Hole. [Online] Wired, October 24, 2012. http://www.wired.com/threatlevel/2012/10/d kim-vulnerability-widespread/all/.
65. **Orlando, Michael.** Vulnerability Note VU#268267: DomainKeys Identified Mail (DKIM) Verifiers may inappropriately convey message trust. [Online] U.S. Department of Homeland Security's United States Computer Emergency Readiness Team (US-CERT), October 24, 2012. http://www.kb.cert.org/vuls/id/268267.
66. **US-CERT.** Report Phishing Sites. [Online] U.S. Department of Homeland Security's United States Computer Emergency Readiness Team (US-CERT). [Cited: January 22, 2013.] http://www.us-cert.gov/nav/report_phishing.html.
67. **The Google Safe Browsing Team.** Report Phishing Page. [Online] Google. [Cited: January 22, 2013.] http://www.google.com/safebrowsing/report_phish/.
68. **Saran, Cliff.** ING Direct implements two-factor authentication. [Online] Computer Weekly, August 17, 2006. http://www.computerweekly.com/news/2240078159/ING-Direct-implements-two-factor-authentication.
69. **Danchev, Dancho.** Attacker: Hacking Sarah Palin's email was easy. [Online] ZDNet, September 18, 2008. http://www.zdnet.com/blog/security/attacker-hacking-sarah-palins-email-was-easy/1939.
70. **Cheng, Roger and McCullagh, Declan.** Yahoo breach: Swiped passwords by the numbers. [Online] CNet, July 12, 2012. http://news.cnet.com/8301-1009_3-57470878-83/yahoo-breach-swiped-passwords-by-the-numbers/.
71. **Microsoft.** Microsoft Security Intelligence Report. [Online] Microsoft, January-June 2012. http://download.microsoft.com/download/C/1/F/C1F6A2B2-F45F-45F7-B788-32D2CCA48D29/Microsoft_Security_Intelligence_Report_Volume_13_English.pdf.
72. **Danchev, Dancho.** Report: Malicious PDF files comprised 80 percent of all exploits for 2009. [Online] ZDNet, February 16, 2010. http://www.zdnet.com/blog/security/report-malicious-pdf-files-comprised-80-percent-of-all-exploits-for-2009/5473.
73. **Perlroth, Nicole.** Department of Homeland Security: Disable Java 'Unless It Is Absolutely Necessary'. [Online] The New York Times, January 14, 2013. http://bits.blogs.nytimes.com/2013/01/14/department-of-homeland-security-disable-java-unless-it-is-absolutely-necessary/.
74. **Dormann, Will.** Vulnerability Note VU#625617: Java 7 fails to restrict access to privileged code. [Online] U.S. Department of Homeland Security's United States Computer Emergency Readiness Team (US-CERT), January 10, 2013. http://www.kb.cert.org/vuls/id/625617.
75. **IC3.** Malware Installed on Travelers' Laptops Through Software Updates on Hotel Internet Connections. [Online] Internet Crime Complaint Center (IC3), May 8, 2012. http://www.ic3.gov/media/2012/120508.aspx.
76. **Keizer, Gregg.** Is Stuxnet the 'best' malware ever? [Online] Computerworld, September 16, 2010. http://www.computerworld.com/s/article/9185919/Is_Stuxnet_the_best_malware_ever_.
77. **Fouda, Amir.** Security Protection. [Online] Microsoft Malware Protection Center, September 7, 2011. http://www.microsoft.com/security/portal/threat/encyclopedia/entry.aspx?Name=Security+Protection.
78. **Gross, Doug.** Virus found in fake Android version of 'Angry Birds: Space'. [Online] CNN, April 12, 2012. http://www.cnn.com/2012/04/12/tech/gaming-gadgets/angry-birds-virus-android/index.html.
79. **Halliday, Derek.** Security Alert: SpamSoldier. [Online] Lookout Mobile Security, December 17, 2012. https://blog.lookout.com/blog/2012/12/17/security-alert-spamsoldier/.
80. **Messmer, Ellen.** Pirated mobile Android and Apple apps getting hacked, cracked and smacked. [Online] Network World, August 20, 2012. http://www.networkworld.com/news/2012/082012-pirated-app-malware-261702.html.
81. **Microsoft Security Intelligence Report.** Deceptive Downloads: Software, Music, and Movies. [Online] Microsoft. [Cited: January 24, 2013.] http://www.microsoft.com/security/sir/story/default.aspx#!deceptive_downloads.

82. **Boscovich, Richard Domingues.** Microsoft Disrupts the Emerging Nitol Botnet Being Spread through an Unsecure Supply Chain. [Online] The Office Microsoft Blog, September 13, 2012. http://blogs.technet.com/b/microsoft_blog/archive/2012/09/13/microsoft-disrupts-the-emerging-nitol-botnet-being-spread-through-an-unsecure-supply-chain.aspx.

83. **Microsoft.** Microsoft Safety Scanner. [Online] Microsoft. [Cited: January 24, 2013.] http://www.microsoft.com/security/scanner/en-us/default.aspx.

84. **Martin, Douglas.** Joybubbles, 58, Peter Pan of Phone Hackers, Dies. [Online] The New York Times, August 20, 2007. http://www.nytimes.com/2007/08/20/us/20engressia.html.

85. **Chen, Brian X.** Get Ready for 1 Billion Smartphones by 2016, Forrester Says. [Online] The New York Times, February 13, 2012. http://bits.blogs.nytimes.com/2012/02/13/get-ready-for-1-billion-smartphones-by-2016-forrester-says/.

86. **Goldman, David.** Your smartphone will (eventually) be hacked. [Online] CNNMoney, September 12, 2012. http://money.cnn.com/2012/09/17/technology/smartphone-cyberattack/index.html.

87. **Norton.** Cybercrime Report 2011. [Online] Symantec Corporation, 2012. http://now-static.norton.com/now/en/pu/images/Promotions/2012/cybercrime/assets/downloads/en-us/NCR-DataSheet.pdf.

88. **Bell, Ian.** Commwarrior.A Virus Targets Cell Phones. [Online] Digital Trends, March 9, 2005. http://www.digitaltrends.com/mobile/commwarriora-virus-targets-cell-phones/.

89. **Gold, Jon.** Researchers reveal new rootkit threat to Android security. [Online] Network World, July 2, 2012. http://www.networkworld.com/news/2012/070212-android-malware-260627.html.

90. **Cowley, Stacy.** NFC exploit. Be very, very careful what your smartphone gets near. [Online] CNNMoney, July 26, 2012. http://money.cnn.com/2012/07/26/technology/nfc-hack/index.htm.

91. **Limer, Eric.** Crazy New Exploit Can Brick Samsung Phones or Steal All Their Data. [Online] Gizmodo, December 16, 2012. http://gizmodo.com/5968879/crazy-new-exploit-can-brick-samsung-phones-or-steal-all-their-data.

92. **Kirk, Jeremy.** Android Botnet Abuses People's Phones for SMS Spam. [Online] CIO, December 17, 2012. http://www.cio.com/article/724237/Android_Botnet_Abuses_People_s_Phones_for_SMS_Spam.

93. **Merica, Dan.** Five things you need to know about U.S. national security. [Online] CNN, July 29, 2012. http://security.blogs.cnn.com/2012/07/29/five-things-you-need-to-know-about-u-s-national-security/.

94. **Goldman, David.** Watching porn is bad for your smartphone. [Online] CNNMoney, February 11, 2013. http://money.cnn.com/2013/02/11/technology/security/smartphone-porn/index.html.

95. **Sengupta, Somini.** U.S. Military Hunts for Safe Smartphones for Soldiers. [Online] The New York Times, June 22, 2012. http://bits.blogs.nytimes.com/2012/06/22/u-s-military-hunts-for-safe-smartphones-for-soldiers/.

96. FCC Smartphone Security Checker. [Online] FCC. [Cited: January 25, 2013.] http://www.fcc.gov/smartphone-security.

97. **Souppourison, Aaron.** iPhone lockscreen can be bypassed with new iOS 6.1 trick. [Online] The Verge, February 14, 2013. http://www.theverge.com/2013/2/14/3987830/ios-6-1-security-flaw-lets-anyone-make-calls-from-your-iphone.

98. **Peterson, Scott.** Exclusive: Iran hijacked US drone, says Iranian engineer (Video). [Online] The Christian Science Monitor, December 15, 2011. http://www.csmonitor.com/World/Middle-East/2011/1215/Exclusive-Iran-hijacked-US-drone-says-Iranian-engineer-Video.

99. **Gallagher, Ryan.** FBI Accused of Dragging Feet on Release of Info About "Stingray" Surveillance Technology. [Online] Slate, October 19, 2012. http://www.slate.com/blogs/future_tense/2012/10/19/stingray_imsi_fbi_accused_by_epic_of_dragging_feet_on_releasing_documents.html.

100. **Sandi National Laboratories.** Sandia builds self-contained, Android-based network to study cyber disruptions and help secure hand-held devices. [Online] Sandi National Laboratories, October 2, 2012. https://share.sandia.gov/news/resources/news_releases/

sandia-builds-self-contained-android-based-network-to-study-cyber-disruptions-and-help-secure-hand-held-devices/.

101. **Perlroth, Nicole.** Hackers in China Attacked The Times for Last 4 Months. [Online] The New York Times, January 30, 2013. http://www.nytimes.com/2013/01/31/technology/chinese-hackers-infiltrate-new-york-times-computers.html.

102. **Symantec.** Symantec Statement Regarding New York Times Cyber Attack. [Online] Symantec, January 31, 2013. http://www.marketwire.com/press-release/symantec-statement-regarding-new-york-times-cyber-attack-nasdaq-symc-1751586.htm.

103. **Goldman, David.** Your antivirus software probably won't prevent a cyberattack. [Online] CNNMoney, January 31, 2013. http://money.cnn.com/2013/01/31/technology/security/antivirus/index.html.

104. **Poeter, Damon.** NBC.com Hacked, Infected With Citadel Trojan. [Online] PC Magazine, February 21, 2013. http://www.pcmag.com/article2/0,2817,2415735,00.asp.

105. **Kelly, Heather.** Is the government doing enough to protect us online? [Online] CNN, July 31, 2012. http://www.cnn.com/2012/07/25/tech/regulating-cybersecurity/index.html.

106. **CNN Political Unit.** Investigation opened into hacked Bush family e-mails. [Online] CNN, February 8, 2013. http://politicalticker.blogs.cnn.com/2013/02/08/investigation-opened-into-hacked-bush-family-emails/.

107. **Gaudin, Sharon.** DARPA chief leaves Pentagon for Google job. [Online] Computerworld, March 13, 2012. http://www.computerworld.com/s/article/9225156/DARPA_chief_leaves_Pentagon_for_Google_job.

108. **Dugan, Regina.** Regina Dugan: From mach-20 glider to humming bird drone. [Online] TED, March 2012. http://www.ted.com/talks/regina_dugan_from_mach_20_glider_to_humming_bird_drone.html.

109. **Segall, Laurie.** Facebook pays $40,000 to bug spotters. [Online] CNNMoney, August 30, 2011. http://money.cnn.com/2011/08/30/technology/facebook_bug_bounty/index.htm.

110. **Facebook.** Bounty. [Online] Facebook. [Cited: January 25, 2013.] http://www.facebook.com/whitehat/bounty/.

111. **Facebook.** White Hats. [Online] Facebook. [Cited: January 25, 2013.] http://www.facebook.com/whitehat/.

112. **Pepitone, Julianne.** Google awards $60,000 prize for Chrome hack. [Online] CNNMoney, October 10, 2012. http://money.cnn.com/2012/10/10/technology/security/google-chrome-hacker-prize/index.html.

113. **Kersey, Jason.** Chrome Releases. [Online] Google, October 10, 2012. http://googlechromereleases.blogspot.com/2012/10/stable-channel-update_6105.html.

Chapter 9
Plan X and Generation Z

> We're the ones who built this Internet. Now we're the ones who have to keep it secure.
> —Army Gen. Keith Alexander, Director of National Security Agency and Commander of U.S. Cyber Command. 2012 Aspen Security Forum (July 27, 2012).

> This is the generation that makes a game out of everything. For them, life is a game.
> —Brian Niccol, Taco Bell's chief marketing and innovation officer (May 2012).

> If you control the code, you control the world. This is the future that awaits us.
> —Marc Goodman, global security advisor and futurist TEDGlobal 2012 (June 28, 2012).

9.1 Plan X: Foundational Cyberwarfare

"Other countries are preparing for a cyber war," said Richard M. George, a former National Security Agency (NSA) cyber security official. "If we're not pushing the envelope in cyber, somebody else will" [1].

Since 2009, the Defense Advanced Research Projects Agency (DARPA) within the U.S. Department of Defense (DoD) has been steadily increasing its cyber research budget to $208 million in fiscal year 2012 [2].

In May 2012, DARPA officially announced Plan X [3]. The Plan X program is explicitly not funding research and development efforts in vulnerability analysis or the generation of cyber weapons. Instead, Plan X will attempt to create revolutionary technologies for understanding, planning, and managing military cyber operations in real-time, large-scale, and dynamic network environments.

N. Lee, *Counterterrorism and Cybersecurity*, DOI: 10.1007/978-1-4614-7205-6_9,   143
© Springer Science+Business Media New York 2013

In November 2012, DARPA issued a call for proposals (DARPA-BAA-13-02) on
Foundational Cyberwarfare (Plan X) [4]:

> Plan X will conduct novel research into the nature of cyberwarfare and support devel-
> opment of fundamental strategies needed to dominate the cyber battlespace. Proposed
> research should investigate innovative approaches that enable revolutionary advances in
> science, devices, or systems. Specifically excluded is research that primarily results in
> evolutionary improvements to the existing state of practice.
>
> The Plan X program seeks to build an end-to-end system that enables the military to
> understand, plan, and manage cyber warfare in real-time, large-scale, and dynamic net-
> work environments. Specifically, the Plan X program seeks to integrate the cyber battles-
> pace concepts of the network map, operational unit, and capability set in the planning,
> execution, and measurement phases of military cyber operations. To achieve this goal, the
> Plan X system will be developed as an open platform architecture for integration with
> government and industry technologies.

9.2 Cyber Battlespace Research and Development

The five-year, $110 million Plan X research program aims to build a prototype
system in five technical areas [1, 4]:

1. System Architecture

The core of the Plan X system is the cyber battlespace graphing engine whose primary task is to receive, store, model, retrieve, and send cyber battlespace information to other Plan X system components. The graphing engine receives real-time
information from various network mapping components and operational overlay
sources. The network mapping components send data that allow the graphing engine
to convert and construct a real-time logical network topology. This information will
include trace route data, link latencies, Border Gateway Protocol (BGP) routes, IP
Time-To-Live (TTL) header analysis, node routing tables, and any other type of
information necessary to assist in constructing the logical network topology. The
Plan X system must be able to model network topologies at Internet-level scales.

Operational execution overlay information is stored as meta-data for each element in the logical network topology. For example, operational overlay information will include the operating system identification, network service profile,
defensive and offensive capabilities, and identification, friend or foe (IFF). The
planning and operational areas will attach another layer of information on top of
this constructed cyber battlespace model. Planning information includes the potential entry nodes, support platform placement, communication paths, and target
sets. Centralizing operational planning and execution status will allow the Plan X
system to show a global heat map of its activities, from conceptual to actual.

Secure software architecture design principles will allow the Plan X system
to operate from Unclassified to Top Secret/Special Compartmented Information/
Special Access Program with the possibility of multiple simultaneous technology
evaluations operating at different security levels and compartments.

2.   Cyber Battlespace Analytics

The primary focus of Cyber Battlespace Analytics is to model, reason, and assist military planners to navigate and build strategically sound and tactically feasible cyber operations. The research areas are twofold: (a) development of automated techniques to assist military planners to construct cyberwarfare plans, and (b) support of wargaming applications, such as modeling opponent moves and counter moves, to optimize planning.

An important goal of Cyber Battlespace Analytics is to understand and quantify cyber battlespace effects including the probabilities of collateral damage. The Plan X system will provide assistance selecting optimal nodes in a cyber battlespace. Node sets might include entry nodes, target nodes, and nodes to avoid.

3.   Mission Construction

Mission Construction develops automated techniques that allow mission planners to graphically construct detailed and robust plans that can be automatically synthesized into an executable mission script. The research involves investigating the structure of cyberwarfare program's control flow graphs (CFGs) and the development of domain specific languages (DSLs).

In a cyberwarfare program, CFG, instructions executed at a node, whether an entry node, support platform, or target node, may transfer program control by "calling" other nodes as the mission progresses. Called nodes execute instructions, returning the calculation results to either the calling node or a central coordination node.

A cyberwarfare DSL will allow operation checkpoints during mission execution for real-time operator interaction, support real-time failover to switch to manual control, enable various levels of autonomous operation, leverage existing formal analysis to detect errors/bugs/inconsistencies, enforce rules of engagement (ROE), and construct a cyber operation "play book" to assist in planning future missions.

4.   Mission Execution

Mission Execution involves research and development in (a) the mission script runtime environment and (b) support platforms.

The mission script runtime environment controls the entire execution of a mission, and supports real-time operator interaction. The runtime environment will use public and commercial toolkits such as Metasploit and Immunity CANVAS to build an extensible API framework for assembling capabilities for each mission program.

Support platforms focus on the development of operating systems and virtual machines designed to execute cyberwarfare missions in highly dynamic and hostile cyber battlespaces. Support platforms include launch platforms, battle effect monitors, communication relay, and adaptive defense support like packet filtering, connection filtering, and mitigation capability.

5.   Intuitive Interfaces

Intuitive Interfaces provide a fully integrated visual user experience for commanders, planners, and operators to manage cyberwarfare activities. Similar to a massively multiplayer online game (MMOG), Plan X will model the cyber battlespace and update it with incoming mapping, operational status, and planning information from potentially millions of users.

Plan X will develop four integrated graphical interface workflows: (a) Real-time cyber battlespace views, (b) Planning process, (c) Capability construction, and (d) Operator controls. Touch user interface (TUI), tablet computing, and augmented reality (AR) displays will be a part of the user interface and user experience of Plan X. Traditional keyboard and mouse interactions will be minimized.

9.3 National Centers of Academic Excellence in Cyber Operations

Unlike DARPA and Plan X, the NSA has been actively involved in building cyber weapons such as Stuxnet and Flame [5]. However, finding skilled employees for cyberwarfare is not easy. "Universities don't want to touch [hacking], they don't want to have the perception of teaching people how to subvert things," said NSA technical director Steven LaFountain [6].

In May 2012, the NSA launched a National Centers of Academic Excellence (CAE) in Cyber Operations Program to prime college students for careers in cyberwarfare. The first four universities to receive the CAE-Cyber Operations designation for the 2012–2013 academic year were Dakota State University, the Naval Postgraduate School, Northeastern University, and the University of Tulsa.

LeFountain remarked on the NSA-funded national centers: "The nation increasingly needs professionals with highly technical cyber skills to help keep America safe today—and to help the country meet future challenges and adapt with greater agility. When it comes to national security, there is no substitute for a dedicated, immensely talented workforce. This effort will sow even more seeds" [7].

9.4 Generation Z, Teen Hackers, and Girl Coders

In addition to recruiting from universities and college graduates, the U.S. government has turned to Generation Z born from the early 2000s to the present day.

Unlike Generation X and Generation Y (Millennial Generation), Generation Z is bestowed with advanced communication and media technology since birth—the Internet, instant messaging, text messaging, smartphones, and tablets, just to name a few. The new generation is sometimes referred to as Generation Wii, iGeneration, Gen Tech, Digital Natives, Net Gen, and Facebook Generation.

Gen Wii, shorthand for connectivity, was coined by Taco Bell executives in consultations with MTV. "This is the generation that makes a game out of everything," said Brian Niccol, chief marketing and innovation officer at Taco Bell. "For them, life is a game" [8].

The stereotype of a gamer or a computer hacker being a bright but socially maladapted adolescent boy is archaic and flawed [9]. In February 2012, Spanish police with support of INTERPOL arrested a 16-year-old girl who was allegedly a member of the Anonymous hacking group [10]. In February 2013, security film AVG linked a piece of malware to an 11-year-old boy in Canada. "We believe these junior programmers are motivated mainly by the thrill of outwitting their peers, rather than financial gain," said AVG's chief technology officer Yuval Ben-Itzhak. "But it is nevertheless a disturbing and increasing trend" [11].

Nevertheless, with the exception of a few, most hackers are not evildoers. Even Facebook CEO and cofounder Mark Zuckerberg calls himself a hacker [12]. "The word 'hacker' has an unfairly negative connotation from being portrayed in the media as people who break into computers," Zuckerberg wrote in the Facebook IPO S-1 filing statement. "In reality, hacking just means building something quickly or testing the boundaries of what can be done" [13].

During his keynote address at the 2012 DEF CON 20 Hacking Conference, NSA director Gen. Keith Alexander brought on stage an 11-year-old girl known by the pseudonym CyFi, and called her "the most important person for our future" [14]

A year before in 2011, DEF CON featured for the very first time a section, dubbed DEFCON Kids, for children ages 8–16. The then 10-year-old CyFi revealed a security flaw and a new class of vulnerability in iPhone and Android games by tinkering with the clock settings on the phone and tablet [15].

Citing the key role of the U.S. government in the early research and development of the Internet, NSA director Alexander hoped to recruit from the cream of the crop at DEF CON: "We're the ones who built this Internet. Now we're the ones who have to keep it secure, and I think you folks can help do that…. In this room, this room right here, is the talent our nation needs to secure cyberspace" [16].

In addition, the NSA careers website in July 2012 dedicated a special page to the conference attendees: "Attention DEF CON® 20 attendees: If you're up on your game, you already know the National Security Agency and what we do…. If you think you saw cool things at DEF CON® 20, just wait until you cross the threshold to NSA, 'cause you ain't seen nothing yet" [17] (see Fig. 9.1).

The NSA is not alone in tapping the Generation Z talent pool. The Central Intelligence Agency (CIA) has been offering "Kids' Page" and "Kids' Games" on the official CIA website since 2007 [18]. Figure 9.2 shows the CIA webpage that offers high-quality children games that rival Disney and Nickelodeon's. "Break the Code" and "Aerial Analysis Challenge" (see Fig. 9.3) are two of the engaging online games and activities from the CIA.

It is an irony that the world's first computer programmer was a female—Ada Lovelace, and yet there have been disproportionately more men than women studying computer science. In May 2012, the National Center for Education Statistics (NCES) released its findings that in 2009–2010, females earned 57 % of all bachelor's degrees, 60 % of all master's degrees, and 52 % of all doctoral degrees [19]. However, in the field of computer and information sciences, only 18 % of undergraduates, 27 % of graduates, and 22 % of doctoral students were women. The biggest drop was the number of bachelor's degrees in computer

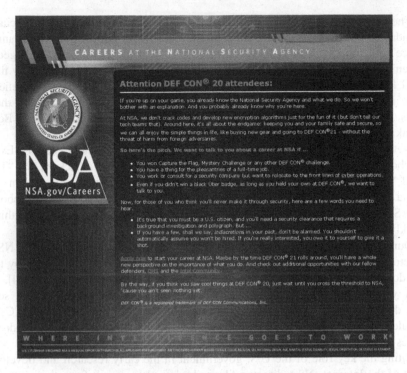

Fig. 9.1   Attention DEF CON 20 attendees (Courtesy of the National Security Agency)

Fig. 9.2   Kids' games on Central Intelligence Agency website (Courtesy of the Central Intelligence Agency)

Fig. 9.3 "Break the Code" and "Aerial Analysis Challenge" games (Courtesy of the Central Intelligence Agency)

science conferred to females, from 28 % in 1999–2000 to 18 % a decade later in 2009–2010.

Generation Z is going to change all that. CyFi is just one example of a girl genius in computer technology. For the first time in history, DEFCON 20 had women show up for the Social Engineering Capture the Flag (SECTF) completion [20]. In 2012, the small number of women hired by game companies has at least tripled since 2009 [21].

Dr. Maria Klawe, president of Harvey Mudd College, and her faculty have achieved the near impossible: nearly 40 % of Harvey Mudd's computer science degrees in 2012 went to women [22]. Moreover, nonprofit educational organizations such as Girls Who Code founded by Reshma Saujani are working to educate, inspire, and equip teenage girls with the skills and resources to pursue opportunities in technology and engineering. Girls Who Code has garnered support from Google, General Electric, eBay, and Twitter. "Our support for this initiative represents our commitment to invest in, encourage and empower more women pursing opportunities in technology," said Dick Costolo, CEO of Twitter [23].

Generation Z women should be encouraged by Virginia Tech engineering alumni Regina Dugan and Letitia "Tish" Long. In 2009, Dugan was appointed the first female director of DARPA. In 2010, Long became the first woman in charge of a major U.S. intelligence agency—the National Geospatial-Intelligence Agency (NGA).

As the interconnected world has become reliant on computers for everything from banking to military, teen hackers and girl coders of Generation Z are powerful resources for the NSA, CIA, DARPA, and U.S. intelligence community to carry out covert cyber operations. NSA director Gen. Keith Alexander spoke at DEF CON 20 in July 2012 in Las Vegas [24]:

> We as a global society are extremely vulnerable and at risk for a catastrophic cyber event. Global society needs the best and brightest to help secure our most valued resources in cyberspace: our intellectual property, our critical infrastructure and our privacy. DEF CON has an important place in computer security. It taps into a broad range of talent and provides an unprecedented diversity of experiences and expertise to solve tough problems. The hacker community and USG cyber community share some core values: we both see the Internet as an immensely positive force; we both believe information increases in value by sharing; we both respect protection of privacy and civil liberties; we both believe in the need for oversight that fosters innovation, doesn't pick winners and losers, and retains freedom and flexibility; we both oppose malicious and criminal behavior. We should build on this common ground because we have a shared responsibility to secure cyberspace.

9.5 Control the Code, Control the World

Bill Gates of Microsoft, Mark Zuckerberg of Facebook, Jack Dorsey of Twitter, will.i.am, and many others appeared in a 2013 YouTube video encouraging everyone to code [25]. Marc Goodman, global security advisor and futurist, spoke at the TEDGlobal 2012 in Edinburgh about his ominous prediction: "If you control the code, you control the world. This is the future that awaits us" [26].

In the 1983 movie *WarGames*, a high school student hacked into a fictional military supercomputer WOPR (War Operation Plan Response) and almost started World War III as a result [27].

In reality, two "war games" were conducted in June and December 2011 between the United States and China in order to help prevent a sudden military escalation between the two countries if either nation felt that they were being targeted [28]. The face-to-face war games were organized by the Center for Strategic and International Studies (CSIS) in Washington and the China Institute of Contemporary International Relations (CICIR) in Beijing.

"We coordinate the war games with the State Department and Department of Defense," said Jim Lewis, senior fellow and director of CSIS. "The officials start out as observers and become participants.... It is very much the same on the Chinese side. Because it is organized between two think tanks they can speak more freely.... The [Chinese officials] who favor cooperation are not as strong as the people who favor conflict" [28].

The world is become increasingly more vulnerable to cyber attacks as we adopt new technologies such as the Automatic Dependent Surveillance—Broadcast system (ADS-B) for air traffic control [29], biometric security systems [30], Apple's iTravel app integration with airport security [31], Internet-controlled door locks and thermostats in home automation [32], embedded Android software in rice cookers and refrigerators [33], and self-driving cars from Toyota and Audi [34].

In March 2010, a disgruntled former employee of a Texas auto center hacked into the company computer and created a bit of havoc with a web-based vehicle-immobilization system. As a result, more than 100 drivers in Austin, Texas found their cars disabled or the horns honking out of control [35].

"We typically don't drive our smartphones at 80 miles an hour," said security strategist Brian Contos at McAfee. "But safety concerns and privacy concerns all culminate when you talk about automobiles. The nightmare scenario is 100 cars on a bridge and 50 % of them hit their brakes and 50 % hit their accelerators. Just the amount of collision that something like that would cause with a remote attack, that's pretty scary stuff" [36].

Setting aside the make-believe of Hollywood, it is conceivable that a future world war will be fought among Generation Z on the Internet and the Intranet, causing global economic meltdown and creating immeasurable damage both online and offline. In 2008, a pseudonymous developer named Satoshi Nakamoto introduced Bitcoin, a peer-to-peer digital currency that bypasses governments and banking systems [37]. Imagine hackers of the future empty all bank accounts, erase all debts, and destroy all backup databases.

In January 2013, Anonymous hacked and defaced the websites of the Massachusetts Institute of Technology (MIT) and the U.S. Sentencing Commission (USSC) in protest over the death of 26-year-old hacktivist Aaron Swartz who succumbed to the pressure of an impending lawsuit and committed suicide earlier in the month [38, 39]. In July 2011, Swartz was charged with breaking into MIT's restricted networks and stealing more than four million academic journal articles from JSTOR [40].

A co-owner of Reddit, Swartz contributed to the original RSS (Really Simple Syndication) specification at the young age of 14. "Aaron is seen as a hero," said Christopher Soghoian, principal technologist and senior policy analyst at the American Civil Liberties Union (ACLU). "He spent a lot of time working to make the Internet a more open place. We lost a really important person who changed the Internet in a positive way, and we all lose out by his departure" [41].

Instead of prosecuting hacktivists like Aaron Swartz, the government and the hacker community should open a dialogue to discuss the broader issues of privacy, intellectual property, security, and civil liberties.

U.S. Navy Admiral and NATO's Supreme Allied Commander James Stavridis said that "instead of building walls to create security, we need to build bridges" [42]. NSA Director and Army General Keith Alexander took the unprecedented first step in reaching out to the hacker community in DEF CON 20 in July 2012 [43].

Control the code, control the world. For the sake of our future, it is the responsibility of Generation X and Generation Y to lead Generation Z into peace rather than war.

References

1. **Nakashima, Ellen.** With Plan X, Pentagon seeks to spread U.S. military might to cyberspace. [Online] The Washington Post, May 30, 2012. http://articles.washingtonpost.com/2012-05-30/world/35458424_1_cyberwarfare-cyberspace-pentagon-agency.
2. **DARPA.** DARPA Director speaks of Offensive Capabilities in Cyber Security. [Online] Defense Advanced Research Projects Agency, March 12, 2012. http://www.darpa.mil/NewsEvents/Releases/2012/03/12c.aspx.
3. —Cyber Experts Engage on DARPA's Plan X. [Online] Defense Advanced Research Projects Agency, October 17, 2012. http://www.darpa.mil/NewsEvents/Releases/2012/10/17.aspx.
4. **Defense Advanced Research Projects Agency.** DARPA-BAA-13-02: Foundational Cyberwarfare (Plan X). [Online] Federal Business Opportunities, November 21, 2012. https://www.fbo.gov/utils/view?id=49be462164f948384d455587f00abf19.
5. **Sanger, David E.** Obama Order Sped Up Wave of Cyberattacks Against Iran. [Online] The New York Times, June 1, 2012. http://www.nytimes.com/2012/06/01/world/middleeast/obama-ordered-wave-of-cyberattacks-against-iran.html?pagewanted=all.
6. **Koebler, Jason.** NSA Built Stuxnet, but Real Trick Is Building Crew of Hackers. [Online] US News and World Report, June 8, 2012. http://www.usnews.com/news/articles/2012/06/08/nsa-built-stuxnet-but-real-trick-is-building-crew-of-hackers.
7. **NSA Public and Media Affairs.** NSA Announces New Program to Prime College Students for Careers in Cyber Ops. [Online] National Security Agency, May 21, 2012. http://www.nsa.gov/public_info/press_room/2012/new_college_cyber_ops_program.shtml.
8. **Horovitz, Bruce.** After Gen X, Millennials, what should next generation be? [Online] USA Today, May 4, 2012. http://usatoday30.usatoday.com/money/advertising/story/2012-05-03/naming-the-next-generation/54737518/1.
9. **WGBH educational foundation.** Studying the psychology of virus writers and hackers: An interview with researcher Sarah Gordon. [Online] PBS Frontline. [Cited: January 4, 2013.] http://www.pbs.org/wgbh/pages/frontline/shows/hackers/whoare/psycho.html.
10. **Whiteman, Hilary.** Interpol arrests suspected 'Anonymous' hackers. [Online] CNN, February 29, 2012. http://www.cnn.com/2012/02/29/world/europe/anonymous-arrests-hacking/index.html.
11. **Dunn, John E.** AVG finds 11 year-old creating malware to steal game passwords. [Online] TechWorld, February 8, 2013. http://news.techworld.com/security/3425185/avg-finds-11-year-old-creating-malware-steal-game-passwords/.
12. **Ortutay, Barbara.** For Facebook 'Hacker Way' is way of life. [Online] Yahoo! News, February 4, 2012. http://news.yahoo.com/facebook-hacker-way–way-life-150559696.html.
13. **Facebook, Inc.** Form S-1. Registration Statement. [Online] U.S. Securities and Exchange Commission, February 1, 2012. http://www.sec.gov/Archives/edgar/data/1326801/000119312512034517/d287954ds1.htm.
14. **Kelly, Heather.** Computer hacking for 8-year-olds. [Online] CNN, July 31, 2012. http://www.cnn.com/2012/07/31/tech/web/def-con-kids-2012/index.html.
15. **Rosenblatt, Seth.** 10-year-old hacker finds zero-day flaw in games. [Online] CNet, August 7, 2011. http://download.cnet.com/8301-2007_4-20089152-12/10-year-old-hacker-finds-zero-day-flaw-in-games/.

16. **Cowley, Stacy.** NSA wants to hire hackers. [Online] CNNMoney, July 29, 2012. http://money.cnn.com/2012/07/27/technology/defcon-nsa/index.htm.
17. **National Security Agency.** Careers. [Online] National Security Agency. http://www.nsa.gov/careers/dc20/.
18. **Central Intelligence Agency.** Games. [Online] Central Intelligence Agency, March 6, 2007. https://www.cia.gov/kids-page/games/index.html.
19. **Aud, Susan; Hussar, William; Johnson, Frank; Kena, Grace; Roth, Erin; Manning, Eileen; Wang, Xiaolei; Zhang, Jijun; Notter, Liz; Nachazel, Thomas; Yohn, Carolyn.** The Condition of Education 2012. [Online] U.S. Department of Education, May 2012. http://nces.ed.gov/pubs2012/2012045.pdf.
20. **Hadnagy, Christopher J. and Maxwell, Eric.** Defcon 20 Social Engineering CTF Report. [Online] DEFCON, September 24, 2012. http://www.social-engineer.org/social-engineering/defcon-20-social-engineering-ctf-report/.
21. **Nayak, Malathi.** Women pry open door to video game industry's boys' club. [Online] Reuters, January 13, 2013. http://www.reuters.com/article/2013/01/13/us-videogames-women-idUSBRE90C0CI20130113.
22. **Hafner, Katie.** Giving Women the Access Code. [Online] The New York Times, April 2, 2012. http://www.nytimes.com/2012/04/03/science/giving-women-the-access-code.html?pagewanted=all&_r=0.
23. **Girls Who Code.** About Girls Who Code. [Online] Girls Who Code. [Cited: January 4, 2013.] http://www.girlswhocode.com/about/.
24. **Alexander, Keith.** Shared Values, Shared Responsibility. [Online] DEF CON Communications Inc., July 2012. https://www.defcon.org/html/defcon-20/dc-20-speakers.html.
25. **Code.org.** What most schools don't teach. [Online] YouTube, February 26, 2013. https://www.youtube.com/watch?v=nKIu9yen5nc
26. **Goodman, Marc.** Marc Goodman: A vision of crimes in the future. [Online] TEDGlobal 2012, June 28, 2012. http://www.ted.com/talks/marc_goodman_a_vision_of_crimes_in_the_future.html?quote=1769.
27. **IMDb.** WarGames. [Online] IMDb, June 3, 1983. http://www.imdb.com/title/tt0086567/.
28. **Hopkins, Nick.** US and China engage in cyber war games. [Online] The Guardian, April 16, 2012. http://www.guardian.co.uk/technology/2012/apr/16/us-china-cyber-war-games.
29. **Kelly, Heather.** Researcher: New air traffic control system is hackable. [Online] CNN, July 26, 2012. http://www.cnn.com/2012/07/26/tech/web/air-traffic-control-security/index.html.
30. **Goldman, David.** Hackers' next target: Your eyeballs. [Online] CNNMoney, July 26, 2012. http://money.cnn.com/2012/07/26/technology/iris-hacking/index.htm.
31. **Patterson, Thom.** Apple's secret plan to join iPhones with airport security. [Online] CNN, September 19, 2012. http://www.cnn.com/2012/09/19/travel/mobile-airport-travel-apps/index.html.
32. **Goldman, David.** Your future home is vulnerable to cyberattacks. [Online] CNNMoney, July 26, 2012. http://money.cnn.com/2012/07/26/technology/home-network-cyberattack/index.htm.
33. **Edwards, Cliff and King, Ian.** Google Android Baked Into Rice Cookers in Move Past Phone. [Online] Bloomberg, January 7, 2013. http://www.bloomberg.com/news/2013-01-08/google-android-baked-into-rice-cookers-in-move-past-phones-tech.html.
34. **Cohen, Adam.** Will Self-Driving Cars Change the Rules of the Road? [Online] Time Magazine, January 14, 2013. http://ideas.time.com/2013/01/14/will-self-driving-cars-change-the-rules-of-the-road/.
35. **Poulsen, Kevin.** Hacker Disables More Than 100 Cars Remotely. [Online] Wired, March 17, 2010. http://www.wired.com/threatlevel/2010/03/hacker-bricks-cars/.
36. **Neild, Barry.** Could hackers seize control of your car? [Online] CNN, March 2, 2012. http://www.cnn.com/2012/03/02/tech/mobile/mobile-car-hacking/index.html.
37. **Nakamoto, Satoshi.** Bitcoin P2P e-cash paper. [Online] GMANE Newsgroup, October 31, 2008. http://article.gmane.org/gmane.comp.encryption.general/12588/
38. **Musil, Steven.** Anonymous hacks MIT after Aaron Swartz's suicide. [Online] CNet, January 13, 2013. http://news.cnet.com/8301-1023_3-57563752-93/anonymous-hacks-mit-after-aaron-swartzs-suicide/.

39. **Brumfield, Ben.** Anonymous threatens Justice Department over hacktivist death. [Online] CNN, January 27, 2013. http://www.cnn.com/2013/01/26/tech/anonymous-threat/index.html.
40. **Seidman,   Bianca.** Internet activist charged with hacking into MIT network. [Online] PBS, July 22, 2011. http://www.pbs.org/wnet/need-to-know/the-daily-need/internet-activist-charged-with-hacking-into-mit-network/.
41. **Leopold, Todd.** How Aaron Swartz helped build the Internet. [Online] CNN, January 15, 2013. http://www.cnn.com/2013/01/15/tech/web/aaron-swartz-internet/index.html.
42. **Stavridis, James.** James Stavridis: A Navy Admiral's thoughts on global security. [Online] TED,   June   2012.   http://www.ted.com/talks/james_stavridis_how_nato_s_supreme_commander_thinks_about_global_security.html.
43. **Urquhart, Conal.** US National Security Agency boss asks hackers to make internet more secure. [Online] The Guardian, July 28, 2012. http://www.guardian.co.uk/technology/2012/jul/28/national-security-agency-hackers-internet.

Part IV
Counterterrorism Strategies: Causes and Cures, War and Peace

Chapter 10
Understanding Terrorism

I study the future of crime and terrorism. And quite frankly I'm afraid. … We consistently underestimate what criminals and terrorists can do
—Marc Goodman, Global security advisor and futurist.
TEDGlobal 2012 (June 28, 2012).

You can kill a man, but you can't kill an idea
—Medgar Evers.
American civil rights activist and U.S. Army Sergeant.

Why kill a man when you can kill an idea?
—Abu Nazir in *Homeland* (TV series).

Morality intervenes not when we pretend we have no enemies but when we try to understand them, to put ourselves in their situation. … Trying to understand other people means destroying the stereotype without denying or ignoring the otherness.
—Umberto Eco (September 4, 2012).

10.1 Bravery and Cowardice

One of my all-time favorite television shows was Bill Maher's *Politically Incorrect* on ABC between 1997 and 2002. On September 11, 2001, former assistant U.S. attorney Barbara Olson was traveling to a *Politically Incorrect* taping in Los Angeles. She was a passenger on American Airlines Flight 77 that crashed into the Pentagon. Maher left a panel chair empty for a week in honor of her memory.

On the September 17 show, Maher's guest Dinesh D'Souza quarreled with President George W. Bush's characterization of the terrorists as cowards. D'Souza argued, "Look at what they [the 9/11 terrorists] did. You have a whole bunch of guys who were willing to give their life; none of them backed out. All of them slammed themselves into pieces of concrete. These are warriors." Maher

N. Lee, *Counterterrorism and Cybersecurity*, DOI: 10.1007/978-1-4614-7205-6_10,   157
© Springer Science+Business Media New York 2013

concurred, "We have been the cowards, lobbing cruise missiles from 2,000 miles away. That's cowardly. Staying in the airplane when it hits the building, say what you want about it, it's not cowardly" [1].

It was shocking, as Maher and his guests often upset the status quo by offering their opposing viewpoints on controversial topics, and sparking heated debates on the show. Maher later apologized and clarified that his "cowardly" comments were directed not at American soldiers, but at "the government, the elected officials, [and] the people who want to put up a giant missile shield, when plainly that's not where the threat is from" [2].

Murdering innocent people who are defenseless is a cowardly act. Standing up to difficult life circumstances and oppressive regimes is bravery. Nevertheless, Maher challenged his millions of viewers, including myself, to dig deeper into the problem of terrorism, rather than fixating on a simplistic good guy versus bad guy paradigm.

In the novel *The Silence of the Lambs*, cannibalistic serial killer Hannibal Lecter told FBI agent-in-training Clarice Starling, "Nothing happened to me. I happened." Like the cold-blooded murderer Lecter, terrorists do not commit evil deeds simply because of some mental disorders or self-centered vendetta. A memorable dialogue between Lecter and Starling follows:

> Lecter: *"First principles, Clarice. Simplicity. Read Marcus Aurelius. Of each particular thing ask: what is it in itself? What is its nature? What does he do, this man you seek?"*
> Starling: *"He kills women…"*
> Lecter: "No. That is incidental. What is the first and principal thing he does? What needs does he serve by killing?"

To combat terrorism, we need to understand terrorism.

10.2 Drones Kill Terrorists, Not Terrorism

Unmanned Aerial Vehicles (UAVs), Unmanned Aerial Systems (UAS), or simply drones have become the counterterrorism weapon of choice. In 2001, the U.S. Department of Defense had fewer than 50 military drones. In 2012, the number of armed and reconnaissance drones swelled to around 7,500 [3]. The Defense Advanced Research Projects Agency (DARPA) has developed a miniature hummingbird-size drone and the Air Force Research Laboratory has plans for a pigeon-sized UAV that can recharge while perching on power lines [4].

The drone program began in the Bush administration to target al-Qaeda leaders in the tribal regions of Pakistan, and it has expanded extensively under President Barack Obama [5]. *The Economist* reported in October 2011 that "under Barack Obama, the frequency of drone strikes on terrorists in Pakistan's tribal areas has risen tenfold, from one every 40 days during George Bush's presidency to one every four" [6].

In January 2012 during a YouTube-Google forum, President Obama defended the use of drones: "I think we have to be judicious in how we use drones.

But understand that probably our ability to respect the sovereignty of other countries and to limit our incursions into somebody else's territory is enhanced by the fact that we are able to (execute a) pinpoint strike on al-Qaeda operatives in a place where the capacities of that military in that country may not be able to get them" [7].

John Brennan, President Obama's assistant for homeland security and counter-terrorism, further explained that "in full accordance with the law—and in order to prevent terrorist attacks on the United States and to save American lives—the United States government conducts targeted strikes against specific al-Qaeda terrorists, sometimes using remotely piloted aircraft, often referred to publicly as drones" [8].

"The drone strikes are not just important in terms of eliminating the leadership of al-Qaeda," said Peter Bergen, director of the national security studies program at the New American Foundation. "They are also important in terms of preventing people from training in the tribal region and making that very difficult because you are always looking over your shoulder for a drone attack" [9].

When American-born al-Qaeda terrorist Anwar al-Awlaki was killed by a drone strike in Yemen in September 2011, a senior U.S. military official said, "It's critically important to send an important message to the surviving leaders and foot soldiers in the Qaeda affiliate. It sets a sense of doom for the rest of them. Getting Awlaki, given his tight operational security, increases the sense of fear. It's hard for them to attack when they're trying to protect their own back side" [10].

However, what is the unintended message that the U.S. is sending to the innocent people residing in their home countries? Many Pakistanis consider U.S. drone strikes a violation of sovereignty and a cause of unacceptable civilian casualties.

According to data compiled by the New America Foundation, "337 CIA [Central Intelligence Agency] drone strikes in Pakistan have killed an estimated 1,932–3,176 people between 2004 and 2012, of which 1,487–2,595 were reported to be militants. This means the average non-militant casualty rate over the life of the program is 18–23 %" [11]. Reports of collateral damage and the civilian reactions to the drone strikes included [12]:

1. 28-year-old Pakistani Mohammad Rehman Khan lost his father, three brothers, and a nephew in South Waziristan on the Afghan border. After Obama's reelection in 2012, Khan said, "The same person who attacked my home has gotten reelected. … When the Sandy hurricane came, I thought that Allah would wipe away America. America just wants to take over the world."
2. Haji Abdul Jabar who lost his 23-year-old son said, "Whenever he has a chance, Obama will bite Muslims like a snake. Look at how many people he has killed with drone attacks."
3. Warshameen Jaan Haji who lost his wife said, "Any American, whether Obama or Mitt Romney, is cruel. I lost my wife in the drone attack and my children are injured. Whatever happens, it will be bad for Muslims."

To raise public awareness of drone attacks, British artist James Bridle launched Dronestagram on Instagram in October 2012 to show Google Earth images of the locations of drone strikes [13]. Bridle told *CNN*, "We have gotten better at immediacy and intimacy online. Perhaps we can be better at empathy too" [14].

Widespread resentment, anger, and hatred toward Americans will not help the U.S. win the war on terror. On the contrary, as *Reuters* reported in November 2012, "anger over the unmanned aircraft may have helped the Taliban gain recruits, complicating efforts to stabilize the unruly border region between Pakistan and Afghanistan. That could also hinder Obama's plan to withdraw U.S. troops from Afghanis" [12].

Indeed, NATO's International Security Assistance Force (ISAF) in Afghanistan released dismal data in September 2012 showing that the conditions in Afghanistan are mostly worse than before the U.S. troop surge began two years ago in 2010 [15]. President Barack Obama had high hopes when he outlined his plan for an Afghanistan troop surge in 2009 [16], but the surge ended in 2012 with little fanfare [17].

American civil rights activist and U.S. Army Sergeant Medgar Evers once said that "you can kill a man, but you can't kill an idea" [18]. The statement applies to both good and evil. Drones kill terrorists, but not the idea of terrorism.

10.3 War on Terror

In the TV series *Homeland*, Abu Nazir told U.S. Marine Sergeant Nicholas Brody, "Why kill a man when you can kill an idea?" [19]. It was a real irony because Nazir played a wanted terrorist leader who converted Brody into an undercover operative for al-Qaeda. Brody was turned after a U.S. drone strike killed Nazir's young child whom Brody grew to love like a son.

In order to win the war on terror, we need to understand terrorism so that we can find ways to end it. The 9/11 Commission Report stated: "The problem is that al-Qaeda represents an ideological movement, not a finite group of people. It initiates and inspires, even if it no longer directs. In this way it has transformed itself into a decentralized force. ... Killing or capturing him [Osama bin Laden], while extremely important, would not end terror. His message of inspiration to a new generation of terrorists would continue" [20].

Wayne Murphy, deputy director for analysis and production at the National Security Agency (NSA) had also said, "In the end, I don't know if the benefits of getting bin Laden would balance out. And I don't know if it buys us anything. Think about what we just went through with Saddam Hussein" [21].

Nevertheless, General Stanley McChrystal and Michael Scheuer represented the majority opinion that bin Laden was a top priority in the war on terror. General McChrystal, then commander of U.S. Forces Afghanistan (USFOR-A), testified in Congress, "I don't think that we can finally defeat al-Qaeda until he's captured or killed. I believe he is an iconic figure at this point, whose survival emboldens al-Qaeda as a franchising organization across the world" [22].

In fact, after the bombings of the American embassies in East Africa, al-Qaeda conspirator Abu Jandal told an FBI agent, "Can you imagine how many joined bin Laden after the embassy bombings? Hundreds came and asked to be martyrs" [23].

Scheuer, a 22-year veteran with the CIA, created and served as the chief of the agency's Osama bin Laden unit at the Counterterrorist Center. Scheuer openly blamed President Bill Clinton's refusal to authorize the CIA to kidnap or assassinate bin Laden [24].

On May 2, 2011, bin Laden was finally found and killed by U.S. Navy SEALs in Abbottabad, Pakistan [25]. By the time of his death, bin Laden was no longer an effective leader but simply a figurehead of the al-Qaeda terrorist organization. He hid in an unguarded compound with no intrusion alarm and no escape route; he was basically a sitting duck waiting to die a martyr or a coward depending on one's point of view.

"Jihad against America will not stop with the death of Osama," proclaimed Fazal Mohammad Baraich, a Muslim cleric. "Osama bin Laden is a shaheed (martyr). The blood of Osama will give birth to thousands of other Osamas" [26].

American-born al-Qaeda terrorist Anwar al-Awlaki was one of "the other Osamas." Dubbed the "bin Laden of the Internet," al-Awlaki had a blog, a Facebook page, and YouTube videos [27]. He had evaded capture for years until September 2011 when he was killed by a CIA drone strike in Yemen. Once again, he was considered a martyr by the Islamic extremists.

"The death of Sheik Anwar al-Awlaki will merely motivate the Muslim youth to struggle harder against the enemies of Islam and Muslims," said Anjem Choudhry, an Islamic scholar in London. "I would say his death has made him more popular" [10].

In spite of bin Laden and al-Awlaki's death, the war on terror continues without an end in sight. Tom Engelhardt of tomdispatch.com wrote, "Unless we set aside the special ops assaults and the drone wars and take a chance, unless we're willing to follow the example of all those nonviolent demonstrators across the Greater Middle East and begin a genuine and speedy withdrawal from the Af/Pak theater of operations, Osama bin Laden will never die" [28].

10.4 A Stubborn Terror

"The American public is underestimating the Islamic fundamentalist groups, and terrorism and extremism," said Amrullah Saleh, head of the National Directorate of Security, Afghanistan's domestic intelligence agency [29].

Bruce Riedel, former CIA officer and current senior fellow at the Brookings Institution, chaired the strategic review of policy toward Afghanistan and Pakistan in 2009 at President Barack Obama's request. On September 10, 2012, a day before the terrorist attack at the U.S. Consulate in Benghazi, Libya, Bruce Riedel published a chilling report titled "A Stubborn Terror" in *Newsweek/The Daily Beast* [30]:

> Eleven years after 9/11, al-Qaeda is fighting back. Despite a focused and concerted American-led global effort—despite the blows inflicted on it by drones, SEALS, and spies—the terror group is thriving in the Arab world, thanks to the revolutions that swept across it in the last 18 months. And the group remains intent on striking inside America and Europe. The al-Qaeda core in Pakistan has suffered the most from the vigorous blows orchestrated by the Obama administration. The loss of Osama bin Laden eliminated its most charismatic leader, and the drones have killed many of his most able lieutenants. But even with all these blows, bin Laden's successor, Ayman al-Zawahiri, is still orchestrating a global terror network and communicating with its followers.

Specifically, Riedel pointed out that:

1. The Taliban and Lashkar-e-Taiba, al-Qaeda's allies in Pakistan, have a global network with terrorist cells in the U.S., England, and the Persian Gulf.
2. Al-Qaeda is multiplying in the Arabian Peninsula, especially Yemen.
3. Al-Qaeda carries out waves of bombings every month in Iraq.
4. Al-Qaeda and other Islamist extremists has taken over half of Mali in North Africa.
5. A new al-Qaeda franchise has emerged in Egypt's Sinai Peninsula.
6. Al-Qaeda operation is fast growing in Syria.
7. Al-Qaeda dispatched Chechen terrorists to Spain in 2012 to attack Gibraltar.
8. Al-Qaeda has planned to attack New York, Chicago, and Detroit since 2009.

U.S. Secretary of State Hillary Clinton testified at the Benghazi hearings before Congress in January 2013: "The Arab Spring has ushered in a time when al-Qaeda is on the rise. The world in many ways is even more dangerous because we lack a central command [in al-Qaeda] and have instead these nodes that are scattered throughout North Africa and other places" [31].

Earlier in January, Islamist militants, including al-Qaeda in the Islamic Maghreb (AQIM), attacked a convoy of oil workers in Algeria and held captive dozens of hostages from the United States, United Kingdom, France, Japan, Norway, and other countries [32].

Besides international terrorism, the U.S. is also facing domestic terrorism by militant right-wing, left-wing, anarchist, and special interest groups.

According to Southern Poverty Law Center (SPLC), the number of conspiracy-minded anti-government "Patriot" groups reached an all-time high of 1,360 in 2012, up 7 % from 2011 [33].

Years before the 9/11 attacks, former U.S. Army platoon leader Timothy McVeigh killed 168 people including 19 children in the Oklahoma City bombing in April 1995 [34]. It was the worst act of homegrown domestic terrorism in U.S. history. During the 1996 Summer Olympics, U.S. Army veteran Eric Rudolph planted a pipe bomb at the Centennial Olympic Park in Atlanta, Georgia, killing 2 and injuring 112 people [35]. In 1996 alone, the Federal Bureau of Investigation (FBI) thwarted five planned acts of domestic terrorism directed against local law enforcement officials in Montana, an FBI facility in West Virginia, communications and transportation infrastructure, and banking facilities in Washington State.

The FBI said, "As long as violence is viewed by some as a viable means to attain political and social goals, extremists will engage in terrorism" [35]. Marc Goodman, global security advisor and futurist, painted a bleak future as he told the audience at TEDGlobal 2012 in Edinburgh: "I study the future of crime and terrorism. And quite frankly I'm afraid. … We consistently underestimate what criminals and terrorists can do" [36].

In December 2012, Dr. Mathew Burrows of the National Intelligence Council (NIC) published a 166-page report, "Global Trends 2030: Alternative Worlds," in which he expressed a mixed outlook on the future of terrorism [37]:

[Positive:] Several circumstances are ending the current Islamist phase of terrorism, which suggest that as with other terrorist waves—the Anarchists in the 1880s and 1890s, the postwar anti-colonial terrorist movements, the New Left in 1970s—the recent religious wave is receding and could end by 2030.

[Negative:] Taking a global perspective, future terrorists could come from many different religions, including Christianity and Hinduism. Right-wing and left-wing ideological groups—some of the oldest users of terrorist tactics— also will pose threats. ... The worst-case outcome on nuclear proliferation ... terrorists or extremist elements acquiring WMD (Weapon of Mass Destruction) material.

[Conclusion:] Terrorism is unlikely to die completely, however, because it has no single cause. The traditional use of the term "root cause" for understanding what drives terrorism is misleading. Rather, some experts point to the analogy of a forest fire: a mixture of conditions—such as dry heat, a spark, and wind—that lead to terrorism.

10.5 Al-Qaeda's Battle is Economic, Not Military

In the December 2012 NIC report "Global Trends 2030: Alternative Worlds," Burrows wrote that "terrorists ... would focus less on causing mass casualties and more on creating widespread economic and financial disruptions" [37]. Al-Qaeda, has in fact, been employing the economic warfare tactic for many years.

In a 2004 videotape sent to Alijazeera, Osama bin Laden spoke of "having experience in using guerrilla warfare and the War of Attrition to fight tyrannical superpowers" as al-Qaeda "bled Russia for 10 years, until it went bankrupt and was forced to withdraw in defeat" and so al-Qaeda is "continuing this policy in bleeding America to the point of bankruptcy" [38].

In October 2005, Abu Mus'ab al-Najadi, a Saudi supporter and member of al-Qaeda authored a seven-page document titled "Al-Qaeda's Battle is Economic not Military" in which -he clarified bin Laden's strategy [39]:

Usually, wars are based on military strength and victory belongs to those who are militarily superior on the battlefield ... But our war with America is fundamentally different, for the first priority is defeating it economically. For that, anything that negatively affects its economy is considered for us a step in the right direction on the path to victory. Military defeats do not greatly effect how we measure total victory, but these defeats indirectly affect the economy which can be demonstrated by the breaching of the confidence of capitalists and investors in this nation's ability to safeguard their various trade and dealings.

This reveals the importance of the blessed September 11th attacks, which is not that it killed a large number of infidels, but what is more important, the economic effect that this strike achieved. ... I will not be exaggerating if I say that striking the Pentagon was purely symbolic and had no noticeable effect on the course of the battle.

The U.S. seems to have fallen into al-Qaeda's trap, as Chris Hedges, former foreign correspondent for *The New York Times*, wrote in February 2011 [40]:

We are wasting $700 million a day to pay for the wars in Iraq and Afghanistan, while our teachers, firefighters and police lose their jobs, while we slash basic assistance programs for the poor, children and the elderly, while we turn our backs on the some 3 million people being pushed from their homes by foreclosures and bank repossessions and while we do nothing to help the one in six American workers who cannot find work. ... [These wars] have turned our nation into an isolated pariah, fueling the very terrorism we seek to defeat. And they cannot be won. The sooner we leave Iraq and Afghanistan the sooner we will save others and finally save ourselves.

The war in Iraq officially ended on December 18, 2011 after a costly, nearly nine-year military engagement. Almost 4,500 U.S. troops had been killed in Iraq since 2003 [41]. The war in Afghanistan began on October 7, 2001 and it has become the longest war in American history, surpassing Iraq (8 years and 7 months) and Vietnam (8 years and 5 months) [42].

Retired Lt. Gen. Robert Gard and Brig. Gen. John Johns wrote a CNN article in December 2012, "In the last decade, America fought two expensive wars and Congress has yet to pay for them; that policy has contributed to our precarious economic position. … Cutting Pentagon spending recognizes that national security is more than military power. The United States is stronger with a strong economy, sustainable jobs, investment in education, renewal of our infrastructure and a sensible energy strategy. Continuing to waste money when our nation should have other priorities is bad policy and bad for security" [43].

In December 2012, U.S. Treasury Secretary Timothy Geithner said that the federal government had already reached the debt ceiling of nearly $16.4 trillion [44].

10.6 Inside the Terrorist Mind

Jeffrey Swanson, a professor in psychiatry and behavioral sciences at Duke University's School of Medicine, debunked a common misconception about terrorists being social misfits afflicted with psychological disorders. Swanson explained that "based on the best available scientific evidence on the link between violence and mental illness in populations, most violence is not caused by a major psychiatric condition like schizophrenia, bipolar disorder, or depression" [45]. Terrorists are different from those sociopathic shooters at Columbine High School, Virginia Tech, Aurora Colorado Century 16 movie theater, or Sandy Hook Elementary School.

On the contrary, Colonel Philip G. Wasielewski reported in *Joint Force Quarterly* that "an analysis of over 150 al-Qaeda terrorists displayed a norm of middle- to upper-class, highly educated, married, middle-aged men" [46]. Women are also appearing in increasing numbers, and have been significant actors in rebel groups such as the Tamil Tigers in Sri Lanka [47]. Even adolescents and children are being manipulated by adults to join armed conflicts in war and terrorism as child soldiers [48]. Kids as young as eight have been used as bombers in Pakistan [49].

To understand terrorism is to comprehend the terrorist mind. Terrorists want their voice heard at the expense of innocent lives including men, women, and children. Terrorists are often self-righteous and motivated by ideologies, politics, religions, and vengeance, either singly or in combination:

1. **Osama bin Laden** was once considered an American ally during the Soviet war in Afghanistan between December 1979 and February 1989. Not only did bin Laden and his jihad fighters receive American and Saudi funding, some analysts believe that bin Laden himself had security training from the CIA

[50]. In January 1991, the U.S. led a coalition force in Operation Desert Storm against Iraq in response to Iraq's invasion of Kuwait [51]. Bin Laden became furious, especially about the continuing U.S. military presence in his birthplace Saudi Arabia long after the Persian Gulf War ended in February 1991. In his 1996 fatwā "Declaration of War against the Americans Occupying the Land of the Two Holy Places," bin Laden wrote [52]:

Terrorizing you, while you are carrying arms on our land, is a legitimate and morally demanded duty. ... our problem will be how to restrain our youths to wait for their turn in fighting and in operations. ... They stood up tall to defend the religion; at the time when the government misled the prominent scholars and tricked them into issuing fatwās of opening the land of the two Holy Places for the Christians armies and handing the Al-Aqsa Mosque to the Zionists. The youths hold you responsible for all of the killings and evictions of the Muslims and the violation of the sanctities, carried out by your Zionist brothers in Lebanon; you openly supplied them with arms and finance. More than 600,000 Iraqi children have died due to lack of food and medicine and as a result of the unjustifiable aggression (sanction) imposed on Iraq and its nation. The children of Iraq are our children. You, the USA, together with the Saudi regime are responsible for the shedding of the blood of these innocent children.

2. **Timothy McVeigh**, former U.S. Army platoon leader who once served in the Persian Gulf, bombed the Murrah Federal Building in Oklahoma City in 1995 on the second anniversary of the Waco siege on April 19 [34]. To justify his criminal act that caused the death of 168 people including 19 children, McVeigh authored "An Essay on Hypocrisy" from a federal maximum-security prison and he wrote a letter to *Fox News* in April 2001 [53]:

I chose to bomb a federal building because such an action served more purposes than other options. Foremost, the bombing was a retaliatory strike; a counter attack, for the cumulative raids (and subsequent violence and damage) that federal agents had participated in over the preceding years (including, but not limited to, Waco.) ... our government—like the Chinese—was deploying tanks against its own citizens. ... Bombing the Murrah Federal Building was morally and strategically equivalent to the U.S. hitting a government building in Serbia, Iraq, or other nations.

3. **Eric Rudolph**, a U.S. Army veteran and self-described Catholic, conducted a series of terrorist acts across southeastern U.S. between 1996 and 1998. He confessed to the 1996 Centennial Olympic Park bombing in Atlanta as well as bombings at a gay nightclub and two abortion clinics [54]. Michael Barkun, professor emeritus of political science at Syracuse University, called Rudolph a "Christian terrorist" [55]. In his April 2005 statement [56] and letters to his mother from behind bars [57], Rudolph offered his reasons:

In the summer of 1996, the world converged upon Atlanta for the Olympic Games. Under the protection and auspices of the regime in Washington millions of people came to celebrate the ideals of global socialism. ... even though the purpose of the Olympics is to promote these despicable ideals, the purpose of the attack on July 27th was to confound, anger and embarrass the Washington government in the eyes of the world for its abominable sanctioning of abortion on demand.

However wrongheaded my tactical decision to resort to violence may have been, morally speaking my actions were justified. ... I don't believe that a government that sanctions, protects and promotes the murder of 50 million unborn children in the heinous practice of abortion has the moral authority to judge a man for murder.

4. **Anders Behring Breivik**, a Norwegian native and self-proclaimed Christian, bombed government buildings in Oslo and shot 85 people dead at a youth camp held by the ruling Labor party on Utoeya Island in July 2011 [58]. Before carrying out the hideous acts, Breivik posted on Facebook to his 7,000+ friends a massive 1,516-page manifesto titled "2083: A European Declaration of Independence" [59], in which he plagiarized the Unabomber manifesto "Industrial Society and Its Future" by American terrorist Ted Kaczynski [60]. Breivik espoused a right-wing, anti-Islam, and anti-immigrant ideology [61]:

Let's end the stupid support for the Palestinians that the Eurabians have encouraged, and start supporting our cultural cousin, Israel. ... I believe Europe should strive for: A cultural conservative approach where monoculturalism, moral, the nuclear family, a free market, support for Israel and our Christian cousins of the east, law and order and Christendom itself must be central aspects (unlike now). Islam must be re-classified as a political ideology and the Quran and the Hadith banned as the genocidal political tools they are.

5. **Anwar al-Awlaki**, an American-born al-Qaeda cleric, was one of the most wanted terrorists. After a CIA drone strike killed al-Awlaki in Yemen in September 2011, his widow **Aminah**, a white European Muslim convert, vowed revenge on the United States. Aminah said, "I would be making a martyr operation, but Sheikh Basir al-Wuhayshi, the emir of AQAP [al-Qaeda in the Arabian Peninsula], said that the sisters so far [can] not carry out operations because it will mean a lot of problems for them ... so I cannot perform operation. ... I want to be killed the same way as my husband was ... Insha'Allah" [62].

Aminah is one among many terrorist sympathizers. Um Saad, a middle-aged woman in the Sunni district of Khadra in west Baghdad, lost her husband and two of her sons in a series of unfortunate events during "the surge" in Iraq in 2007. Saad exclaimed, "[The Americans] are monsters and devils wearing human clothes. One day I will put on an explosive belt under my clothes and then blow myself up among the Americans. I will get revenge against them for my husband and sons and I will go to paradise [63]."

Ed Husain, senior fellow for Middle Eastern studies at the Council on Foreign Relations (CFR), wrote an opinion piece on *CNN* in November 2012: "Al-Jazeera Arabic gives prominence to the popular Egyptian Muslim Brotherhood cleric Yusuf al-Qaradawi, who has repeatedly called suicide bombings against Israelis not terrorism, but 'martyrdom.' He argues that since Israelis all serve in the military, they are not civilians. Even children, he despicably argues, are not innocent. They would grow up to serve in the military. Qaradawi is not alone. I can name tens of Muslim clerics, important formulators of public opinion in a region dominated by religion, that will readily condemn acts of terrorism against the West, but will fall silent when it comes to condemning Hamas or Islamic Jihad. Put simply, support for violent resistance against Israel among Arab and Muslim-majority countries—including allies of the United States such as Qatar, Saudi Arabia, Egypt, Turkey, Tunisia—remains popular" [64].

10.7 Questioning Terrorism and Destroying Stereotypes

Paul R. Pillar, former national intelligence officer for the Near East and South Asia, reported that there has been a global decline of state-sponsored international terrorism since the end of the Cold War [65]. However, MIT Professor Emeritus Noam Chomsky opines that the United States is a leading terrorist state. In a November 2001 interview by David Barsamian, Chomsky cited Nicaragua as an example: In 1984, the U.S. was "the only country that was condemned for international terrorism by the World Court and that rejected a Security Council resolution calling on states to observe international law" [66].

Going back in history to the 1950s, *The New York Times* published in April 2000 a CIA document written in 1954 on "Operation Ajax" (aka TPAJAX)—a successful plot to overthrow the democratically-elected Iranian Prime Minister Mohammad Mossadegh. The document shows that "Iranians working for the CIA and posing as Communists harassed religious leaders and staged the bombing of one cleric's home in a campaign to turn the country's Islamic religious community against Mossadegh's government" [67].

Chomsky also blamed the Reagan administration for the 1985 Beirut car bombing near the home of Mohammad Hussein Fadlallah in a failed assassination attempt that instead killed about 60 and injured more than 200 civilians [68]. If the allegation was true, the administration would be in direct violation of President Reagan's Executive Order 12333 stating that "no person employed by or acting on behalf of the United States Government shall engage in or conspire to engage in assassination" [69].

Two years prior in 1983, the Beirut barracks bombing killed 220 U.S. Marines, 21 Americans, and 58 French paratroopers [70]. The United States categorized the bombing an act of terrorism towards off-duty American servicemen, but the Hezbollah regarded the Americans and Frenchmen as enemy combatants stationing in Beirut.

Fast forward to the post 9/11 era, one might argue that CIA drone attacks are state-sponsored assassinations. However, the U.S. Congress approved the Authorization for the Use of Military Force (AUMF), giving the President the authority to use "all necessary and appropriate force against those nations, organizations, or persons he determines planned, authorized, committed, or aided the terrorist attacks that occurred on September 11, 2001, or harbored such organizations or persons, in order to prevent any future acts of international terrorism against the United States by such nations, organizations or persons" [71].

Collateral damages from CIA drone strikes are pale in comparison to economic sanctions. In May 1996, CBS reporter and former White House correspondent Lesley Stahl asked U.S. Secretary of State Madeleine Albright on *60 Minutes*, "We have heard that half a million children have died [during the U.N. sanctions against Iraq]. I mean, that's more children than died in Hiroshima. And—and you know, is the price worth it?" Albright replied, "I think this is a very hard choice, but the price—we think the price is worth it" [72]. In her autobiography *Madam Secretary: A Memoir*, Albright regretted making that statement and she wrote, "My reply had been a terrible mistake... Nothing matters more than the lives of innocent people" [73].

After the 9/11 terrorist attacks, *The New York Times* reported in September 2001 that the Bush administration demanded Pakistan to cut off fuel supplies and eliminate truck convoys that provide much of the food and medicines to Afghanistan's civilian population [74].

Chomsky considered the U.S. sanctions and many CIA covert operations to be equivalent to state-sponsored terrorism: "The U.S. is officially committed to what is called 'low-intensity warfare.' That's the official doctrine. If you read the definition of low-intensity conflict in Army manuals and compare it with official definitions of 'terrorism' in Army manuals, or the U.S. Code, you find they're almost the same. Terrorism is the use of coercive means aimed at civilian populations in an effort to achieve political, religious, or other aims" [66].

HBO's *Real Time with Bill Maher* host wrote in his blog on November 30, 2012: "We utilize the best means at our disposal to go into foreign lands and blow up the people we consider the bad guys even if that means collateral damage in the form of civilian casualties. When someone does that exact same thing to us, don't we call it 'terrorism'?" [75].

Italian philosopher Umberto Eco said, "I would argue that morality intervenes not when we pretend we have no enemies but when we try to understand them, to put ourselves in their situation. ... Trying to understand other people means destroying the stereotype without denying or ignoring the otherness" [76].

References

1. **Bohlen, Celestine.** In New War on Terrorism, Words Are Weapons, Too. [Online] The New York Times, September 29, 2001. http://www.nytimes.com/2001/09/29/arts/think-tank-in-new-war-on-terrorism-words-are-weapons-too.html.
2. **ABC News.** Maher Apologizes for 'Cowards' Remark. [Online] ABC News, September 20, 2001. http://abcnews.go.com/Entertainment/story?id=102318&page=1.
3. **Bergen, Peter and Rowland, Jennifer.** A dangerous new world of drones. [Online] CNN, October 8, 2012. http://www.cnn.com/2012/10/01/opinion/bergen-world-of-drones/index.html.
4. **Weinberger, Sharon.** Pentagon's Tiny New Spy Drone Mimics Hummingbird. [Online] AOL News, February 18, 2011. http://www.aolnews.com/2011/02/18/pentagons-tiny-new-spy-drone-mimics-hummingbird/.
5. **Levine, Adam.** Obama admits to Pakistan drone strikes. [Online] CNN, January 30, 2012. http://security.blogs.cnn.com/2012/01/30/obama-admits-to-pakistan-drone-strikes/.

6. **The Economist.** Flight of the drones: Why the future of air power belongs to unmanned systems. [Online] The Economist, October 8, 2011. http://www.economist.com/node/21531433.

7. **Jackson, David.** Obama defends drone strikes. [Online] USA Today, January 31, 2012. http://content.usatoday.com/communities/theoval/post/2012/01/obama-defends-drone-strikes/1.

8. **Kelly, Suzanne.** Deadly drones and the classified conundrum. [Online] CNN, May 23, 2012. http://security.blogs.cnn.com/2012/05/23/deadly-drones-and-the-classified-conundrum/.

9. **Benson, Pam.** Yemen plot exposes new world of U.S. spying. [Online] CNN, May 11, 2012. http://security.blogs.cnn.com/2012/05/11/yemen-plot-exposes-new-world-of-us-spying/.

10. **Mazzetti, Mark, Schmitt, Eric and Worth, Robert F.** Two-Year Manhunt Led to Killing of Awlaki in Yemen. [Online] The New York Times, September 30, 2011. http://www.nytimes.com/2011/10/01/world/middleeast/anwar-al-awlaki-is-killed-in-yemen.html?pagewanted=all.

11. **New America Foundation.** An Analysis of U.S. Drone Strikes in Pakistan, 2004-2012. [Online] New America Foundation. [Cited: December 2, 2012.] http://counterterrorism.newamerica.net/drones.

12. **Fabi, Randy and Chowdhry, Aisha.** Obama victory infuriates Pakistani drone victims. [Online] Chicago Tribune, November 8, 2012. http://articles.chicagotribune.com/2012-11-08/news/sns-rt-us-usa-campaign-pakistanbre8a70a0-20121107_1_drone-strikes-drone-program-pakistani-taliban.

13. **Bridle, James.** Dronestagram: The drone's-eye view. [Online] Instagram, October 2012. http://instagram.com/dronestagram.

14. **Boyette, Chris.** Dronestagram uses social media to highlight drone strikes. [Online] CNN, February 15, 2013. http://www.cnn.com/2013/02/14/tech/dronestagram/index.html.

15. **Ackerman, Spencer.** Military's Own Report Card Gives Afghan Surge an F. [Online] Wired, September 27, 2012. http://www.wired.com/dangerroom/2012/09/surge-report-card/.

16. **Obama, Barack.** Obama Outlines Plan for Afghanistan Troop Surge. [Online] PBS Newshour, December 1, 2009. http://www.pbs.org/newshour/updates/military/july-dec09/obamapeech_12-01.html.

17. **Nordland, Rod.** Troop 'Surge' in Afghanistan Ends With Mixed Results. [Online] The New York Times, September 21, 2012. http://www.nytimes.com/2012/09/22/world/asia/us-troop-surge-in-afghanistan-ends.html?pagewanted=all.

18. **Arlington National Cemetery.** Medgar Wiley Evers. [Online] Arlington National Cemetery Website, October 11, 2009. http://www.arlingtoncemetery.net/mwevers.htm.

19. **Haglund, David and Thomas, June.** The Riveting Season Finale of Homeland. [Online] Slate, December 19, 2011. http://www.slate.com/blogs/browbeat/2011/12/19/homeland_season_finale_discussing_the_dramatic_end_to_the_showtime_show_s_first_season_.html.

20. **The 9-11 Commission.** The 9/11 Commission Report: Final Report of the National Commission on Terrorist Attacks Upon the United States (9/11 Report). [Online] U.S. Congress, July 22, 2004. http://www.9-11commission.gov/report/index.htm.

21. **Wright, Lawrence.** The Spymaster. Can Mike McConnell fix America's intelligence community? [Online] The New Yorker, January 21, 2008. http://www.newyorker.com/reporting/2008/01/21/080121fa_fact_wright?currentPage=all.

22. **BBC.** Gen McChrystal: Bin Laden is key to al-Qaeda defeat. [Online] BBC News, December 9, 2009. http://news.bbc.co.uk/2/hi/americas/8402138.stm.

23. **Wright, Lawrence.** The Agent. Did the C.I.A. stop an F.B.I. detective from preventing 9/11? [Online] The New Yorker, July 10, 2006. http://www.newyorker.com/archive/2006/07/10/060710fa_fact_wright?currentPage=all.

24. **Scheuer, Michael F.** Bill and Dick, Osama and Sandy. [Online] The Washington Times, July 4, 2006. http://www.washingtontimes.com/news/2006/jul/4/20060704-110004-4280r/.

25. **Phillips, Macon.** Osama Bin Laden Dead. [Online] The White House Blog, May 2, 2011. http://www.whitehouse.gov/blog/2011/05/02/osama-bin-laden-dead.

26. **msnbc.com staff and news service reports.** Protesters condemn 'brutal killing' of bin Laden. [Online] NBCNews.com, May 6, 2011. http://www.msnbc.msn.com/id/42927020/ns/world_news-death_of_bin_laden/t/protesters-condemn-brutal-killing-bin-laden/.

27. **Madhani, Aamer.** Cleric al-Awlaki dubbed 'bin Laden of the Internet'. [Online] USA Today, August 24, 2010. http://usatoday30.usatoday.com/news/nation/2010-08-25-1A_Awlaki25_C V_N.htm.

28. **Engelhardt, Tom.** Osama Bin Laden's American Legacy: It's Time to Stop Celebrating and Go Back to Kansas. [Online] The Huffington Post, May 5, 2011. http://www.huffingtonpost.com/ tom-engelhardt/osama-bin-ladens-american_b_858182.html.

29. **CBS News.** Ex-CIA Operative Comes Out of the Shadows. [Online] CBS News, August 2, 2010. http://www.cbsnews.com/8301-18560_162-6014887.html?pageNum=2.

30. **Riedel, Bruce.** A Stubborn Terror. [Online] Newsweek/The Daily Beast, September 10, 2012. http://www.thedailybeast.com/newsweek/2012/09/09/a-stubborn-terror.html.

31. **Dougherty, Jill and Cohen, Tom.** Clinton takes on Benghazi critics, warns of more security threats. [Online] CNN, January 24, 2013. http://www.cnn.com/2013/01/23/politics/ clinton-benghazi/index.html.

32. **Fleishman, Jeffrey.** Algeria hostages reportedly escape captors; some may have been slain. [Online] Los Angeles Times, January 17, 2013. http://www.latimes.com/news/nationworld/ world/la-20-foreign-hostages-escape-islamist-captors-in-algeria-20130117,0,7004240.story.

33. **Potok, Mark.** The Year in Hate and Extremism. Intelligence Report, Issue Number 149. [Online] Southern Poverty Law Center, Spring 2013. http://www.splcenter.org/home/2013/ spring/the-year-in-hate-and-extremism

34. **Federal Bureau of Investigation.** Terror Hits Home: The Oklahoma City Bombing. [Online] Federal Bureau of Investigation. [Cited: December 1, 2012.] http://www.fbi.gov/about-us/ history/famous-cases/oklahoma-city-bombing.

35. **Counterterrorism Threat Assessment and Warning Unit, National Security Division.** Terrorism in the United States 1996. [Online] Federal Bureau of Investigation, 1996. http://www.fbi.gov/stats-services/publications/terror_96.pdf.

36. **Goodman, Marc.** The technological future of crime: Marc Goodman at TEDGlobal 2012. [Online] TED, June 28, 2012. http://blog.ted.com/2012/06/28/ the-technological-future-of-crime-marc-goodman-at-tedglobal-2012/.

37. **National Intelligence Council.** Global Trends 2030: Alternative Worlds. [Online] U.S. National Intelligence Council, December 2012. http://www.dni.gov/files/documents/GlobalTr ends_2030.pdf.

38. **Bin Laden, Osama.** Full transcript of bin Ladin's speech. [Online] Aljazeera, November 1, 2004. http://www.aljazeera.com/archive/2004/11/200849163336457223.html.

39. **Salama, Sammy and Wheeler, David.** From the Horse's Mouth: Unraveling Al-Qa`ida's Target Selection Calculus. [Online] James Martin Center for Nonproliferation Studies (CNS), April 17, 2007. http://cns.miis.edu/stories/070417.htm.

40. **Hedges, Chris.** No Other Way Out. [Online] Common Dreams, February 28, 2011. http://www.commondreams.org/view/2011/02/28-0.

41. **Cutler, David.** Timeline: Invasion, surge, withdrawal; U.S. forces in Iraq. [Online] Reuters, December 18, 2011. http://www.reuters.com/article/2011/12/18/ us-iraq-usa-pullout-idUSTRE7BH08E20111218.

42. **Dermody, William.** The Longest War. [Online] USA Today. http://www.usatoday.com/news/ afghanistan-ten-years-of-war/index.html.

43. **Gard, Robert G. and Johns, John.** Generals: Get real and cut Pentagon spending. [Online] CNN, December 12, 2012. http://www.cnn.com/2012/12/12/opinion/gard-johns-military-spending/ index.html.

44. **CNN Political Unit.** Treasury Department rules out $1 trillion coin. [Online] CNN, January 12, 2013. http://politicalticker.blogs.cnn.com/2013/01/12/ treasury-department-rules-out-1-trillion-coin/.

45. **Swanson, Jeffrey.** Looking into the minds of killers. [Online] CNN, July 25, 2012. http://ww w.cnn.com/2012/07/24/opinion/swanson-colorado-shooting/index.html.

46. **Wasielewski, Philip G.** Defining the War on Terror. [Online] Joint Force Quarterly, Issue 44, 1st Quarter 2007, pp. 13-18, 2007. http://www.ndu.edu/press/lib/pdf/jfq-44/JFQ-44.pdf.

47. **Murray, Rebecca.** Scarred by Sri Lanka's war with Tamil Tigers, female ex-fighters build new lives. [Online] The Christian Science Monitor, October 29, 2010. http://www.csmonitor.com/World/Asia-South-Central/2010/1029/Scarred-by-Sri-Lanka-s-war-with-Tamil-Tigers-female-ex-fighters-build-new-lives.
48. **Glazer, Ilsa M.** Armies of the Young: Child Soldiers in War and Terrorism (review). [Online] Anthropological Quarterly, Volume 79, Number 2, Spring 2006, pp. 373-384, 2006. http://dx.doi.org/10.1353%2Fanq.2006.0021.
49. **Mohsin, Saima; Khan, Shaan.** Police: Kids young as 8 used as bombers in Pakistan. [Online] CNN, March 14, 2013. http://www.cnn.com/2013/03/14/world/asia/pakistan-child-bombers/index.html
50. **BBC News.** Al-Qaeda's origins and links. [Online] BBC News, July 20, 2004. http://news.bbc.co.uk/2/hi/middle_east/1670089.stm.
51. **Chief of Naval Operations.** The United States Navy In "Desert Shield"/"Desert Storm". [Online] Naval History & Heritage, May 15, 1991. http://www.history.navy.mil/wars/dstorm/index.html.
52. **Public Broadcasting Service.** Bin Laden's Fatwa. [Online] PBS Newshour, August 23, 1996. http://www.pbs.org/newshour/updates/military/july-dec96/fatwa_1996.html.
53. **McVeigh, Timothy.** McVeigh's Apr. 26 Letter to Fox News. [Online] Fox News, April 26, 2001. http://www.foxnews.com/story/0,2933,17500,00.html.
54. **Fonda, Daren.** How Luck Ran Out For A Most Wanted Fugitive. [Online] Time Magazine, June 9, 2003. http://www.time.com/time/magazine/article/0,9171,1004966,00.html.
55. **Olsen, Ted.** Is Eric Rudolph a Christian Terrorist? [Online] Christianity Today, June 1, 2003. http://www.christianitytoday.com/ct/2003/juneweb-only/6-2-22.0.html.
56. **Rudolph, Eric.** Full text of Eric Rudolph's written statement. [Online] Army of God, April 13, 2005. http://www.armyofgod.com/EricRudolphStatement.html.
57. **Morrison, Blake.** Special report: Eric Rudolph writes home. [Online] USA Today, July 5, 2005. http://usatoday30.usatoday.com/news/nation/2005-07-05-rudolph-cover-partone_x.htm.
58. **BBC.** Norway police say 85 killed in island youth camp attack. [Online] BBC News, July 23, 2011. http://www.bbc.co.uk/news/world-europe-14259356.
59. **Boston, William.** Killer's Manifesto: The Politics Behind the Norway Slaughter. [Online] Time Magazine, July 24, 2011. http://www.time.com/time/world/article/0,8599,2084901,00.html.
60. **Breivik, Anders Behring.** Behring Breivik kopierte Una-bomberen. [Online] DOCUMENT. no, July 24, 2011. http://www.document.no/2011/07/behring-breivik-kopierte-una-bomberen/.
61. **Kane, Alex.** Breivik manifesto outlines virulent right-wing ideology that fueled Norway massacre. [Online] Mondoweiss, July 24, 2011. http://mondoweiss.net/2011/07/breivik-manifesto-outlines-virulent-right-wing-ideology-that-fueled-norway-massacre.html.
62. **Cruickshank, Paul, Lister, Tim and Robertson, Nic.** The Danish agent, the Croatian blonde and the CIA plot to get al-Awlaki. [Online] CNN, October 24, 2012. http://www.cnn.com/2012/10/15/world/al-qaeda-cia-marriage-plot/index.html.
63. **Cockburn, Patrick.** Bereaved Iraqi mother vows revenge on US. [Online] The Independent, March 13, 2008. http://www.independent.co.uk/news/world/middle-east/bereaved-iraqi-mother-vows-revenge-on-us-795018.html.
64. **Husain, Ed.** Israel, face new reality: Talk to Hamas. [Online] CNN, November 21, 2012. http://www.cnn.com/2012/11/20/opinion/husain-hamas-israel/index.html.
65. **Pillar, Paul R.** The Decline of State-Sponsored Terrorism. [Online] The Atlantic, May 22, 2012. http://www.theatlantic.com/international/archive/2012/05/the-decline-of-state-sponsored-terrorism/257515/.
66. **Chomsky, Noam.** The United States is a Leading Terrorist State. [Online] Monthly Review, November 2001. http://monthlyreview.org/2001/11/01/the-united-states-is-a-leading-terrorist-state.
67. **Risen, James.** Secrets of History: The C.I.A. in Iran. [Online] The New York Times, April 2000. http://www.nytimes.com/library/world/mideast/041600iran-cia-index.html.
68. **The Guardian.** 60 killed by Beirut car bomb. [Online] The Guardian, March 9, 1985. http://www.guardian.co.uk/theguardian/1985/mar/09/fromthearchive.

69. **Office of the President of the United States.** Executive Order 12333: United States Intelligence Activities. [Online] Central Intelligence Agency, December 4, 1981. https://www.cia.gov/about-cia/eo12333.html#2.11.
70. **Hampson, Rick.** 25 years later, bombing in Beirut still resonates. [Online] USA Today, October 18, 2008. http://usatoday30.usatoday.com/news/military/2008-10-15-beirut-barracks_N.htm.
71. **107th Congress Public Law 40.** Authorization for the Use of Military Force (AUMF). [Online] U.S. Government Printing Office, September 18, 2001. http://www.gpo.gov/fdsys/pkg/PLAW-107publ40/html/PLAW-107publ40.htm.
72. **David, Leigh and Wilson, James.** Counting Iraq's victims. [Online] The Gudardian, October 10, 2001. http://www.guardian.co.uk/world/2001/oct/10/iraq.socialsciences.
73. **Albright, Madeleine Korbel and Woodward, William.** Madam Secretary: A Memoir. [Online] Miramax Books, April 6, 2005. http://books.google.com/books?id=RBuEq2f5U_QC.
74. **Burns, John F.** AFTER THE ATTACKS: IN ISLAMABAD; Pakistan Antiterror Support Avoids Vow of Military Aid. [Online] The New York Times, September 16, 2001. http://www.nytimes.com/2001/09/16/us/after-attacks-islamabad-pakistan-antiterror-support-avoids-vow-military-aid.html?src=pm&pagewanted=all.
75. **Maher, Bill.** Spacial Delivery. [Online] HBO, November 30, 2012. http://www.real-time-with-bill-maher-blog.com/real-time-with-bill-maher-blog/2012/11/30/spacial-delivery.html.
76. **Eco, Umberto.** Inventing the Enemy: Essays. [Online] Houghton Mifflin Harcourt, September 4, 2012. http://books.google.com/books?id=mFbYcy8uhCUC&pg=PA18&lpg=PA18.

Chapter 11
Cures for Terrorism

> It's a different type of war. Dealing with terror is going to be
> more like managing disease.
> —Henry "Hank" Crumpton, former deputy director of the CIA
> Counterterrorism Center (July 27, 2012).

> Our job as citizens is to ask questions.
> —Thomas Blanton, National Security Archive.
> George Washington University (December 16, 2010).

> We do not have the right to resort to violence—or the threat of
> violence—when we don't get our way.
> —President Bill Clinton (April 18, 2010).

> The tools to change the world are in everybody's hands, and
> how we use them is…up to all of us…. Public safety is too
> important to leave to the professionals.
> —Marc Goodman at TEDGlobal 2012.
> (June 28, 2012).

> Peace is the only path to true security. … No wall is high
> enough, and no Iron Dome is strong enough, to stop every
> enemy from inflicting harm.
> —President Barack Obama (March 21, 2013).

11.1 Terrorism as a Disease

Dracunculiasis, also known as Guinea worm disease (GWD), is caused by the parasite Dracunculus medinensis. The disease affects communities that do not have safe water to drink. There is no vaccine or drug therapy for Guinea worm disease. Through health education and innovative low-cost water treatments, the Carter Center has led the effort to eradicate the disease in collaboration with the Centers for Disease Control and Prevention (CDC), the World Health Organization (WHO),

N. Lee, *Counterterrorism and Cybersecurity*, DOI: 10.1007/978-1-4614-7205-6_11,   173
© Springer Science+Business Media New York 2013

the United Nations Children's Fund (UNICEF), and the Bill & Melinda Gates Foundation. Two decades of eradication efforts have successfully reduced Guinea worm disease infection cases from 3.5 million worldwide in 1986 to a miniscule 542 cases in 2012 [1]. The Carter Center has predicted that "Guinea worm disease is poised to be the next human disease after smallpox to be eradicated" [2].

Henry "Hank" Crumpton, former deputy director of the Central Intelligence Agency (CIA) Counterterrorism Center, led an insurgent to overthrow the Taliban and to attack al-Qaeda in Afghanistan just after 9/11. Crumpton spoke at the Aspen Security Forum in July 2012 about the war on terror: "It's a different type of war. Dealing with terror is going to be more like managing disease" [3].

There are two fundamental ways to manage disease: treat the symptoms or remedy the root cause. Crumpton chose the former, the symptomatic treatment. In his 2010 interview on *60 Minutes*, Crumpton told CBS correspondent Lara Logan, "[My] orders were fairly simple: Find al-Qaeda and kill them, especially leadership. Destroy command and control…. If they kill me, I have told my family and my friends not to complain about anything, because I have killed many of them with pride" [4].

Trying to get rid of the symptoms (terrorists) without paying attention to the root cause (terrorist motives) does not eradicate the disease but may instead exacerbate it. In spite of the operational successes, Crumpton admitted, "There will be an attack in the homeland. And sadly I think we face that prospect in the future. I think we'll be hit again." When Logan asked if such an attack would be on the scale of 9/11, he responded, "It's certainly possible. Or perhaps even greater" [4].

American author and philosopher Henry David Thoreau wrote in his book *Walden; or, Life in the Woods* that "there are a thousand hacking at the branches of evil to one who is striking at the root" [5]. Understanding terrorism is the prerequisite in tackling the deadly disease. The previous chapter of the book examines the root cause of terrorism. This chapter explores the cures for terrorism through education, communication tools, and innovative treatments:

1. "Revenge is sour"—George Orwell
2. "Govern your passions or they will be your undoing"—Mr. Spock
3. "Impossible to carry a grudge and a big dream at the same time"—Unknown
4. "Every truth has two sides"—Aesop
5. "Give everyone a voice"—Mark Zuckerberg
6. "The only security of all is in a free press"—Thomas Jefferson
7. "Free speech would not protect a man falsely shouting fire"—Oliver Wendell Holmes, Jr.
8. "198 methods of nonviolent action"—Gene Sharp
9. "We do not have the right to resort to violence when we don't get our way"—President Bill Clinton
10. "Peace is the only path to true security"—President Barack Obama

At the TEDGlobal 2012 conference in Edinburgh, global security advisor and futurist Marc Goodman said, "The tools to change the world are in everybody's hands, and how we use them is…up to all of us…. Public safety is too important to leave to the professionals" [6].

11.2 "Revenge is Sour": George Orwell

The ancient Code of Hammurabi dates back to 1792-50 B.C. when the Mesopotamian king Hammurabi ruled the Babylonian Empire. Out of the collection of 282 laws, the most famous Hammurabi's Code is "an eye for an eye, and a tooth for a tooth". However, the law of retaliation treated various social classes of society differently. Hammurabi wrote, "If a man has destroyed the eye of a man of the gentleman class, they shall destroy his eye. If he has destroyed the eye of a commoner, he shall pay one mina of silver. If he has destroyed the eye of a gentleman's slave, he shall pay half the slave's price" [7].

Fast forward 3,800 years to the present time, double standards continue to prevail in our modern world. Age, gender, race, ethnicity, religion, politics, sexual orientation, and ideological belief never cease to affect human judgment. A ruling party may label its opponents "terrorists" in order to delegitimize them, but an oppressive regime may commit acts of violence against its own people. From Timothy McVeigh's point of view, the U.S. government was the villain in the Waco siege. He carried out his revenge by the Oklahoma City bombing without realizing that murdering innocent people is not really a punishment for the U.S. government.

In his essay "Revenge is Sour," George Orwell argued that "the whole idea of revenge and punishment is a childish daydream. Properly speaking, there is no such thing as revenge. Revenge is an act which you want to commit when you are powerless and because you are powerless: as soon as the sense of impotence is removed, the desire evaporates also" [8].

Osama bin Laden was powerless in persuading his countryman Saudi King Fahd and Defense Minister Prince Sultan bin Abdul Aziz to deny American intervention in the Persian Gulf War [9]. Codenamed "Operation Desert Storm," the multinational coalition forces from the United States, Saudi Arabia, the United Kingdom, and Egypt liberated Kuwait from the Iraqi invasion in a mere 100 hours after the ground campaign started in February 1991 [10]. Although the U.S. military forces remained in Saudi Arabia after the war to maintain the stability of the region, President George H. W. Bush refused to capture Baghdad and overthrow Saddam Hussein. Meanwhile, Osama bin Laden, powerless to remove the American troops from his home country, resorted to terrorism against the United States in Saudi Arabia, Tanzania, Kenya, Yemen, and eventually New York.

In April 2003, U.S. Defense Secretary Donald Rumsfeld and Saudi Defense Minister Sultan bin Abdul Aziz announced the withdrawal of all U.S. combat forces from Saudi Arabia [11]. It was not because of the 9/11 attacks in September 2001, in which 15 of the 19 hijackers involved were Saudis. Osama bin Laden and terrorism did not win. Instead, the people of the Middle East and North Africa have influenced American foreign policy by voicing their strong disapproval of the U.S. presence in Saudi Arabia.

Reasoning is better than revenge. According to a 2008 Gallup poll, 60 % of Egyptians, 52 % of Saudis, 55 % of Tunisians, and 29 % of Turks agreed that removing all U.S. military bases from Saudi Arabia would significantly improve their opinion of the United States [12]. The relative friendliness of Turks towards Americans allowed the U.S. to deploy 400 American forces and two Patriot missile batteries in Turkey to counter the Syrian threat [13].

Revenge can backfire. On September 11, 2012, Ansar al-Sharia attacked the U.S. Consulate in Benghazi, Libya to avenge the death of Abu Yahya al-Libi, a-Qaeda's second in command killed by a U.S. drone strike in Pakistan [14]. Their vengeance resulted in the death of U.S. ambassador J. Christopher Stevens, a well-liked figure in Libya. Demonstrations erupted in Benghazi and thousands of Libyans stormed the Ansar al-Sharia headquarters, chanting "You terrorists, you cowards. Go back to Afghanistan" [15].

Revenge is sour. In the series finale of the counterterrorism television show *24*, on-screen hero Jack Bauer (played by Kiefer Sutherland) was persuaded by his most trusted ally Chloe O'Brian (played by Mary Lynn Rajskub) to stop taking his own revenge. Sir Francis Bacon wrote in *Essayes and counsels, civil and moral*: "This is certain, that a man that studies revenge, keeps his own wounds green, which otherwise would heal, and do well" [16]. On the contrary, an eye for an eye will only make the whole world blind.

11.3 "Govern Your Passions or They will be Your Undoing": Mr. Spock

Launched by Pope Urban II in November 1095, the series of Crusades against Muslims and pagans ended more than seven hundred years ago in 1291 [17]. If we really can learn from history, violence should no longer be an option for advanced civilization.

"As a matter of cosmic history, it has always been easier to destroy than to create," Mr. Spock told Dr. McCoy in *Star Trek II: The Wrath of Khan*. When McCoy raised the moral question about the terraforming Genesis Device, Spock replied, "You must learn to govern your passions; they will be your undoing" [18].

The undoing of Osama bin Laden was due to his decision to destroy America rather than to create a bridge between two cultures. Imagine how much good bin Laden could have done with his estimated $300 million inheritance and $7 million yearly stipend [19, 20]. He could have used that money to support the Muslim Educational Cultural Center of America (MECCA), the Council on American-Islamic Relations (CAIR), or even his own grassroots campaigns to influence U.S. foreign policy.

What can a die-hard extremist do to relieve his anger and frustration? One of the answers is cage-fighting, also known as mixed-martial arts (MMA).

Since September 2010, many of the 250-some convicted pro al-Qaeda terrorists have been released from UK prison after having served their terms. Jonathan Evans, director general of the British security service MI5, issued a warning, "We know that some of these prisoners are still committed extremists who are likely to return to their terrorist activities" [21].

Abu Bakr Mansha was one of those convicted terrorists and sentenced to jail at the age of 21. After his second release from prison in March 2011, Mansha sought help from Usman Raja—one of the most renowned cage-fighting coaches in the U.K. Since 2010, Raja has successfully rehabilitated over ten released prisoners into mainstream society [22]. Employing cage-fighting sessions and the teachings of Sheikh Aleey Qadir, Raja's MMA gym has de-radicalized terror convicts with a 100 % success rate. "Take away someone's hate and they feel liberated," explained Raja. "The key is to give them a sense of purpose" [23].

Now a transformed man, Mansha tries to prevent other young Muslims following in his footsteps of terror. "I could channel my energy straight away and build something for myself," said Mansha. "My transformation came over time" [23].

Raja and his wife Khadija hope to expand their MMA gym to wean a generation away from the lure of extremism. Khadija Raja told *CNN*, "There is a real problem here and it's growing and it would be incredibly sad that we have the cure and then it's not delivered and it's not given the platform it needs to be dispersed. At the end of the day I'm still a mother and I don't want to live a world where there is this very real fear" [22].

11.4 "Impossible to Carry a Grudge and a Big Dream at the Same Time"

"My biggest fear is that he'll start to think, 'I will find the person who did this to me and I will do the same to him.' He will live in a world of revenge. I don't want this. I don't want to be the father of a terrorist" [24]. Those were the words of the father of a 7-year-old Bangladeshi boy "Okkhoy" who was severely beaten and mutilated by the beggar mafia in late 2010 because he refused to beg.

The 2008 Oscar-winning movie *Slumdog Millionaire* depicted the horror of forced begging in Bombay, India. Some of the real Mumbai street urchins told *The Telegraph* reporter that the violence in the film was, if anything, understated [25]. Indeed, what happened in real life to a 7-year-old boy was an indescribable atrocity: The attackers cracked open Okkhoy's head, sliced his throat and chest, slashed open his stomach, and chopped off his penis and his right testicle.

Against all odds, Okkhoy survived the brutal attack. *CNN* called him "Okkhoy"—the Bengali word for "unbreakable." His real name was withheld for safety reason. Okkhoy never attended school and his only goal in life was "to be a RAB [Rapid Action Battalion] member and nothing else." He said, "When I grow up, I want to bring them [the attackers] to justice." The RAB is known for unwarranted lethal force, illegal torture, and extrajudicial killings [26].

After *CNN* first aired the Okkhoy story in 2011, businessman Aram Kovach in Columbus, Ohio, decided to help the boy with reconstructive surgery in the United States. Okkhoy finally arrived in the U.S. a year later in 2012. At Johns Hopkins Children's Center in Baltimore, Maryland, a medical team led by Dr.

John Gearhart volunteered to perform a penile surgery on Okkhoy. Gearhart told *CNN*, "As far as an injury committed by one person against another, to a child, this is the most severe genital injury that I've ever seen in 23 years of doing this." Fortunately, the operation was a resounding success.

Okkhoy and his father were touched by the kindness of strangers. Before they left Baltimore, Okkhoy changed his life goal from joining RAB to becoming a doctor. He said, "I want to become a doctor, because I want to save people. And when I do, I won't take any money from them."

It has been said that "it's nearly impossible to carry a grudge and a big dream at the same time." Okkhoy would seek justice in court against the attackers, but he would not be consumed by revenge and miss the opportunity to achieve a much bigger dream.

11.5 "Every Truth has Two Sides": Aesop

Ancient Greek fabulist Aesop said, "Every truth has two sides; it is as well to look at both, before we commit ourselves to either."

Unlike on-screen hero Jack Bauer in the counterterrorism TV drama *24*, real-life hero and FBI agent Ali Soufan did not use torture or coercion on the al-Qaeda terrorist Abu Jandal (aka Nasser al-Bahri) in order to extract information from him about the 9/11 attacks [27]. A former chief bodyguard of Osama bin Laden, Jandal considered himself a revolutionary who believed in the radical Islamist view of history that most of the world's evil came from America, a country that he knew practically nothing about.

During interrogation, Soufan brought Jandal a history book on the United States, in the Arabic language. Jandal was amazed to learn of the American Revolution and its struggle against tyranny. Soufan also showed Jandal a local Yemeni newspaper with the headline that read, "Two Hundred Yemeni Souls Perish in New York Attack." Shaken by the horrific atrocity of 9/11, Jandal drew a breath and said, "God help us.... What kind of Muslim would do such a thing?... The Israelis must have committed the attacks on New York and Washington. The Sheikh [Osama bin Laden] is not that crazy" [28].

After a long debate, Soufan persuaded Jandal to identify his associates in a book of mug shots. Jandal acknowledged knowing seven of the al-Qaeda members, but insisted that bin Laden would never commit the 9/11 attacks. Soufan took the seven photographs out of the book and said, "I know for sure that the people who did this were al-Qaeda guys." Jandal asked, "How do you know? Who told you?" Soufan replied, "You did. These *are* the hijackers. You just identified them." Jandal eventually told Soufan everything he knew and sadly declared, "I think the Sheikh went crazy."

There are always two sides to every story, every political system, and every religious belief. Soufan was able to open Jandal's eyes to see both sides of the story, not by torture or coercion, but by education and debate.

In a 2011 interview with Lara Logan in CBS' *60 Minutes*, Soufan said, "You need to connect with people on a human level—regardless, if they don't like you, want to kill you…. They were trained that we are so evil and we torture and we kill and that is the reason of the rage against us. I try to deprive them from [the rage]" [29].

U.S. Navy Admiral James Stavridis, Commander of the U.S. European Command (USEUCOM) and NATO's Supreme Allied Commander Europe (SACEUR), spoke at the TEDGlobal 2012 conference in Edinburgh about teaching Afghan soldiers to read. Stavridis reasoned that "instead of building walls to create security, we need to build bridges" [30].

In December 2012, Jesuit priest and peace activist John Dear went to Kabul to meet the Afghan Peace Volunteers, a diverse community of students ages 15–27 who practice peace and nonviolence [31]. "I used to detest other ethnic groups," one of the youths told Dear, "but now I'm trying to overcome hate and prejudice. You international friends give me hope and strength to do this." Another youth added, "I used to put people in categories and couldn't drink tea with anyone. Now I'm learning that we are all part of one human family. Now I can drink tea with anyone" [32].

The world needs more Ali Soufan and John Dear to encourage people from diverse cultures and religions to look at both sides of truth. As Nigerian novelist Chimamanda Ngozi Adichie told the audience at TED 2009, "The single story creates stereotypes, and the problem with stereotypes is not that they are untrue, but that they are incomplete. They make one story become the only story" [33]. She said that our lives and our cultures are composed of many overlapping stories, and if we hear only a single story about another person or country, we risk a critical misunderstanding.

11.6 "Give Everyone a Voice": Mark Zuckerberg

Airborne leaflets have been used for military propaganda purposes before and after World War II [34]. Pamphlets were air dropped to the enemy's territories to disseminate the other side of the story. With the advent of the Internet and social networks, both sides have the chance to reach the targeted audience or mass public in real time.

As the world is increasingly becoming connected, 75 % of world leaders utilized Twitter in 2012 to engage their citizens and the global community [35]. Their Twitter use was up from 42 % a year ago in 2011. According to *Digital Daya*, the heads of state in the top 10 list are: President Barack Obama of the United States (24.6 million Twitter followers), President Hugo Chávez of Venezuela (3.8), President Abdullah Gül of Turkey (2.6), Queen Rania of Jordan (2.5), President Dmitry Medvedev of Russia (2.1), President Dilma Rousseff of Brazil (1.8), President Cristina Fernández de Kirchner of Argentina (1.5), President Juan Manuel Santos of Colombia (1.5), President Enrique Peña Nieto of Mexico (1.4), and Sheikh Mohammed bin Rashid Al Maktoum of the United Arab Emirates (UAE) and Dubai (1.3).

British Prime Minster David Cameron who once said "too many tweets make a twat" changed his mind and sent out his first tweet in October 2012. Cameron said, "In this modern world you have got to use every means to try and communicate your message and explain to people why you are doing it" [36].

An open internet is an open platform for debating opposing views. It allows both popular and unpopular voices to be heard. It is a civilized outlet for frustrated individuals to express themselves without resorting to violence or terrorism. U.S. Air Force Senior Airman Christopher R. Atkins wrote in his email to American filmmaker Michael Moore, "Every time we voice our opinion we are promoting freedom" [37].

A light-hearted example is a September 2012 YouTube video called *We Are Hungry* created by high school students in Kansas [38]. With more than a million views, the video successfully forced the Obama administration to reverse some of the new rules by allowing more meats and grains in school lunches. "Even though we're a small town in rural western Kansas, Washington did hear us," said Superintendent Dave Porter. "Our concerns were listened to" [39].

In the January 2012 Facebook IPO letter, Mark Zuckerberg wrote [40]:

> At Facebook, we're inspired by technologies that have revolutionized how people spread and consume information. We often talk about inventions like the printing press and the television—by simply making communication more efficient, they led to a complete transformation of many important parts of society. They gave more people a voice. They encouraged progress. They changed the way society was organized. They brought us closer together. Today, our society has reached another tipping point. We live at a moment when the majority of people in the world have access to the internet or mobile phones—the raw tools necessary to start sharing what they're thinking, feeling and doing with whomever they want. Facebook aspires to build the services that give people the power to share and help them once again transform many of our core institutions and industries. *There is a huge need and a huge opportunity to get everyone in the world connected, to give everyone a voice and to help transform society for the future.* The scale of the technology and infrastructure that must be built is unprecedented, and we believe this is the most important problem we can focus on.
>
> ...
>
> We believe building tools to help people share can bring a more honest and transparent dialogue around government that could lead to more direct empowerment of people, more accountability for officials and better solutions to some of the biggest problems of our time. By giving people the power to share, we are starting to see people make their voices heard on a different scale from what has historically been possible. These voices will increase in number and volume. They cannot be ignored. Over time, we expect governments will become more responsive to issues and concerns raised directly by all their people rather than through intermediaries controlled by a select few.

Even terrorists embrace the idea of Facebook. An online jihadist who goes by the name Rakan al-Ashja'i said, "We will benefit from the ideas in Facebook a lot.... If I could make a social networking website with the same capabilities and everything like Facebook when it first appeared—it is a very good idea" [41].

With over one billion active monthly users in 2012, Facebook is in a unique position to influence the world through the billion-strong human rally against violence. Horrified by the Sandy Hook Elementary School massacre, Beth Howard told her Facebook friends on December 14 about heading to Newtown,

Connecticut to hand out free apple pies to the grieving parents and anyone in need. Within two hours, she received $2,000 in donations. After several days of intense baking with the help of more than 60 volunteers, she drove 1,100 miles to Newtown in her RV loaded with 240 apple pies. "They [the volunteers] were making pies for Newtown because of this one Facebook comment," said Howard. "That was a powerful thing" [42].

11.7 "The Only Security of All is in a Free Press": Thomas Jefferson

"The only security of all is in a free press," said President Thomas Jefferson, American Founding Father and principal author of the Declaration of Independence. "The force of public opinion cannot be resisted when permitted freely to be expressed" [43].

After the 9/11 attacks, Bill Maher's comments on *Politically Incorrect* and Susan Sontag's article in *The New Yorker* challenged President George W. Bush's notion of cowardice, resulting in public uproar and backlash [44]. In a statement supporting Maher, ABC said that *Politically Incorrect* is "a show that celebrates freedom of speech and encourages the animated exchange of ideas and opinions. While we remain sensitive to the current climate following last week's tragedy... there needs to remain a forum for the expression of our nation's diverse opinions" [45].

In 1971, *The New York Times* publisher Arthur Ochs "Punch" Sulzberger decided to publish a top-secret government history of the Vietnam War known as the Pentagon Papers [46]. A federal court ordered the newspaper to halt publishing the Pentagon Papers, citing national security concerns. However, the U.S. Supreme Court ruled on First Amendment grounds that publication could resume—a landmark ruling on press freedom. Exposing the Johnson administration's systematic lies to the American public and U.S. Congress, the Pentagon Papers leaked by Daniel Ellsberg helped hasten the end of the U.S. involvement in the Vietnam War as President Richard Nixon began to withdraw American troops from Vietnam.

President Barack Obama praised Sulzberger as "a firm believer in the importance of a free and independent press, one that isn't afraid to seek the truth, hold those in power accountable and tell the stories that need to be told" [47].

However, quote approval has become a standard practice in the Obama and Romney presidential campaigns in 2012. Under the Obama administration, reporters are forbidden to identify or quote the official speakers in a "deep-background briefing" such as the one held after the U.S. Supreme Court's health care ruling in June 2012 [48].

Former CBS news anchor Dan Rather argues that newspapers and media outlets must push back on quote approval because "submitting to these new tactics puts us more in the category of lapdogs." Rather wrote in a July 2012 *CNN* article, "A free and truly independent press—fiercely independent when necessary—is the red beating heart of freedom and democracy. One of the most important roles of our journalists is to be watchdogs" [49].

Journalists should be watchdogs, not lapdogs. In the documentary film *Fahrenheit 9/11*, director Michael Moore accused the American corporate media of being President George W. Bush's "cheerleaders," instead of providing an accurate and objective analysis of the rationale for the 2003 Iraq War to topple Saddam Hussein [50]. The Walt Disney Company blocked its Miramax division from distributing the film due to the concern that it would jeopardize tax breaks Disney was receiving for its theme park business in Florida, where Bush's brother, Jeb Bush, was governor. Moore criticized Disney's decision: "At some point the question has to be asked, 'Should this be happening in a free and open society where the monied interests essentially call the shots regarding the information that the public is allowed to see?'" [51].

After U.S. Army private Bradley Manning was arrested in February 2012 for divulging three-quarters of a million secret documents to WikiLeaks, op-ed columnist Bill Keller of *The New York Times* wrote [52]:

> In the immediate aftermath of the breach, several news organizations (including this one) considered creating secure online drop-boxes for would-be leakers, imagining that new digital Deep Throats would arise. But it now seems clear that the WikiLeaks breach was one of a kind—and that even lesser leaks are harder than ever to come by. Steven Aftergood, who monitors secrecy issues for the Federation of American Scientists, said that since WikiLeaks the government has elevated the 'insider threat' as a priority, and tightened access to classified material. Nudged by an irate Congress, the intelligence agencies are at work on an electronic auditing program that would make illicit transfer of secrets much more difficult and make tracking the leaker much easier.

A strong supporter of WikiLeaks, the Pentagon Papers whistleblower Daniel Ellsberg said, "Julian Assange is not a criminal under the laws of the United States. I was the first one prosecuted for the charges that would be brought against him. I was the first person ever prosecuted for a leak in this country—although there had been a lot of leaks before me. That's because the First Amendment kept us from having an Official Secrets Act…. The founding of this country was based on the principle that the government should not have a say as to what we hear, what we think, and what we read…. We're not in the mess we're in, in the world, because of too many leaks…. I say there should be some secrets. But I also say we invaded Iraq illegally because of a lack of a Bradley Manning at that time" [53].

Thomas Blanton, director of National Security Archive at George Washington University, testified before the U.S. House of Representatives in December 2010 that "our job as citizens is to ask questions," and he said [54]:

> I wish all terrorist groups would write the local U.S. ambassador a few days before they are launching anything—the way Julian Assange wrote Ambassador Louis Susman in London on November 26—to ask for suggestions on how to make sure nobody gets hurt.
>
> I wish all terrorist groups would partner up with *Le Monde* and *El Pais* and *Der Spiegel* and *The Guardian*, and *The New York Times*, and take the guidance of those professional journalists on what bombs go off and when and with what regulators.

Julian Assange said in a 2011 interview, "Our No. 1 enemy is ignorance. And I believe that is the No. 1 enemy for everyone—it's not understanding what actually is going on in the world" [55]. Indeed, the two-way street of Total Information Awareness is the road that leads to a more transparent and complete picture of ourselves, our governments, and our world (see Chap. 6). Former FBI agent Ali Soufan successfully demonstrated that a history book of America in Arabic language was one of the essential tools in disarming a terrorist. The more information that countries and peoples have about each other, the better and safer the world will become.

English author Edward Bulwer-Lytton wrote in *Richelieu, or The conspiracy: in five acts* that "the pen is mightier than the sword" [56]. Through uncensored journalistic investigations and opinion pieces presenting both sides of the coin, the press can eliminate the need for terrorists to commit violent crimes in order to get their messages across.

11.8 "Free Speech Would Not Protect a Man Falsely Shouting Fire": Oliver Wendell Holmes, Jr.

While information is the oxygen of the modern age, disinformation is the carbon monoxide that can poison generations. Oliver Wendell Holmes, Jr. was an Associate Justice of the United States Supreme Court for 30 years from 1902 to 1932. In his "clear and present danger" majority opinion in the 1919 case of *Schenck v. United States*, Holmes stated that "the most stringent protection of free speech would not protect a man in falsely shouting fire in a theatre and causing a panic.... The question in every case is whether the words used are used in such circumstances and are of such a nature as to create a clear and present danger that they will bring about the substantive evils that Congress has a right to prevent" [57].

In 1919, the United States was marred by domestic terrorism in a series of bombings and assassination attempts orchestrated by the American followers of Italian anarchist Luigi Galleani [58]. Violent activities terrorized New York City, Boston, Pittsburgh, Cleveland, Washington D.C., Philadelphia, and other U.S. cities.

In 1969, protection for speech was raised in *Brandenburg v. Ohio* from "clear and present danger" to "imminent lawless action." Clarence Brandenburg, a Ku Klux Klan (KKK) leader in rural Ohio was convicted under the Ohio criminal syndicalism statute for advocating unlawful methods of racist terrorism as a means of accomplishing political reform. However, the U.S. Supreme Court reversed Brandenburg's conviction because "the constitutional guarantees of free speech and free press do not permit a State to forbid or proscribe advocacy of the use of force or of law violation except where such advocacy is directed to inciting or producing imminent lawless action and is likely to incite or produce such action" [59].

In September 2012, a YouTube trailer of an anti-Islamic film by Nakoula Basseley Nakoula (aka Sam Bacile) wreaked chaos in the Middle East. Despite the Islamic prohibition of portraying Prophet Mohammed, Nakoula's low-budget movie *Innocence of Muslims* depicts the prophet as a womanizer, buffoon, ruthless killer, and child molester. As a result, protestors stormed US embassies in Libya and Egypt, and thousands of demonstrators took to the streets in Afghanistan, Indonesia, Pakistan, Yemen, Lebanon, and Iraq [60]. U.S. Secretary of State Hillary Clinton promptly issued a statement on September 11, 2012 that "the United States deplores any intentional effort to denigrate the religious beliefs of others" [61].

Google temporarily blocked Nakoula's YouTube video in Libya and Egypt. "We work hard to create a community everyone can enjoy and which also enables people to express different opinions," said Google in a written statement. "This can be a challenge because what's OK in one country can be offensive elsewhere. This video—which is widely available on the web—is clearly within our guidelines and so will stay on YouTube. However, given the very difficult situation in Libya and Egypt we have temporarily restricted access in both countries" [62].

Al-Qaeda terrorist Anwar al-Awlaki was dubbed the "bin Laden of the Internet." He used a blog, a Facebook page, and YouTube videos to recruit new jihadists [63]. Before his death in 2011, more than 5,000 postings featuring al-Awlaki's videos were available on YouTube. His video sermons inspired British-born Roshonara Choudhry to become the first al-Qaeda sympathizer to attempt a political assassination in the U.K. by stabbing British MP Stephen Timms in May 2010. In response to the British government's complaint, Google said, "Community Guidelines prohibit dangerous or illegal activities such as bomb-making, hate speech, or incitement to commit specific and serious acts of violence.... We have removed a significant number of videos under these policies. We're now looking into the new videos that have been raised with us and will remove all those which break our rules" [64].

The films by Nakoula and al-Awlaki were similar in that the former's is a direct insult on Islam and the latter's is a direct attack on non-Muslims. In a telephone interview with *The Wall Street Journal*, Nakoula said that his film was "a political movie, not a religious movie" [65]. Although Nakoula did not violate any U.S. law, his action was intentional and irresponsible, akin to a man falsely shouting fire in a theatre and causing a panic.

In September 2012, the American Freedom Defense Initiative (AFDI), led by Pamela Geller and Robert Spencer, ran an ad in 10 subway stations around New York that read: "In any war between the civilized man and the savage, support the civilized man. Support Israel. Defeat Jihad" [66]. New York's Metropolitan Transportation Authority (MTA) initially rejected the ad, but a district judge ruled that the ad was protected under the First Amendment. New York mayor Michael Bloomberg defended the court's decision by saying, "Americans tolerate things they might find despicable because of the First Amendment's protection of free expression" [67].

However, many New Yorkers including Rabbi Rachel Kahn-Troster found the ad "deeply misguided and disturbing." Kahn-Troster said, "The words from our mouths have power: Once released, whether intentionally or by accident, what we say shapes reality. It can bring about healing or atonement, or it can unleash violence and hatred. Geller's ads, sharply dividing the world into civilized people and savages, are only intended to hurt and tear fragile relationships apart" [68].

In *The Story of My Experiments with Truth*, Mahatma Gandhi in 1925 warned that "just as an unchained torrent of water submerges whole countrysides and devastates crops, even so an uncontrolled pen serves but to destroy" [69]. There is a fine line between free speech and hate speech. Free speech encourages debate whereas hate speech incites violence. In September 2012, the United Methodist Women funded a counter ad in New York's subway stations. It reads, "Hate speech is not civilized. Support peace in word and deed" [70].

11.9 "198 Methods of Nonviolent Action": Gene Sharp

Political scientist Gene Sharp, nominated for the 2009 Nobel Peace Prize and highly regarded as the father of nonviolent struggle, compiled 198 methods of nonviolent action in his 1973 book *The Politics of Nonviolent Action, Vol. 2: The Methods of Nonviolent Action* [71]. The list of methods is publicly and freely available online at The Albert Einstein Institution: http://www.aeinstein.org/organizations/org/198_methods.pdf [72]. The comprehensive methods are grouped into six major categories and 37 subcategories ranging from individual to group to nationwide to international activities:

I. Nonviolent Protest and Persuasion
   (a) Formal Statements
       1. Public speeches
       2. Letters of opposition or support
       3. Declarations by organizations and institutions
       4. Signed public statements
       5. Declarations of indictment and intention
       6. Group or mass petitions
   (b) Communications with a Wider Audience
       7. Slogans, caricatures, and symbols
       8. Banners, posters, and displayed communications
       9. Leaflets, pamphlets, and books
       10. Newspapers and journals
       11. Records, radio, and television
       12. Skywriting and earthwriting
   (c) Group Representations
       13. Deputations
       14. Mock awards
       15. Group lobbying
       16. Picketing
       17. Mock elections
   (d) Symbolic Public Acts
       18. Displays of flags and symbolic colors
       19. Wearing of symbols
       20. Prayer and worship
       21. Delivering symbolic objects
       22. Protest disrobings
       23. Destruction of own property
       24. Symbolic lights
       25. Displays of portraits
       26. Paint as protest
       27. New signs and names
       28. Symbolic sounds
       29. Symbolic reclamations

        30. Rude gestures
    (e) Pressures on Individuals
        31. "Haunting" officials
        32. Taunting officials
        33. Fraternization
        34. Vigils
    (f) Drama and Music
        35. Humorous skits and pranks
        36. Performances of plays and music
        37. Singing
    (g) Processions
        38. Marches
        39. Parades
        40. Religious processions
        41. Pilgrimages
        42. Motorcades
    (h) Honoring the Dead
        43. Political mourning
        44. Mock funerals
        45. Demonstrative funerals
        46. Homage at burial places
    (i) Public Assemblies
        47. Assemblies of protest or support
        48. Protest meetings
        49. Camouflaged meetings of protest
        50. Teach-ins
    (j) Withdrawal and Renunciation
        51. Walk-outs
        52. Silence
        53. Renouncing honors
        54. Turning one's back
II. Social Noncooperation
    (a) Ostracism of Persons
        55. Social boycott
        56. Selective social boycott
        57. Lysistratic nonaction (aka crossed legs movement or sex strikes)
        58. Excommunication
        59. Interdict
    (b) Noncooperation with Social Events, Customs, and Institutions
        60. Suspension of social and sports activities
        61. Boycott of social affairs
        62. Student strike
        63. Social disobedience
        64. Withdrawal from social institutions
    (c) Withdrawal from the Social System

65. Stay-at-home
66. Total personal noncooperation
67. "Flight" of workers
68. Sanctuary
69. Collective disappearance
70. Protest emigration (hijra)

III. Economic Noncooperation: Boycotts
(a) Actions by Consumers
71. Consumers' boycott
72. Nonconsumption of boycotted goods
73. Policy of austerity
74. Rent withholding
75. Refusal to rent
76. National consumers' boycott
77. International consumers' boycott
(b) Action by Workers and Producers
78. Workmen's boycott
79. Producers' boycott
(c) Action by Middlemen
80. Suppliers' and handlers' boycott
(d) Action by Owners and Management
81. Traders' boycott
82. Refusal to let or sell property
83. Lockout
84. Refusal of industrial assistance
85. Merchants' "general strike"
(e) Action by Holders of Financial Resources
86. Withdrawal of bank deposits
87. Refusal to pay fees, dues, and assessments
88. Refusal to pay debts or interest
89. Severance of funds and credit
90. Revenue refusal
91. Refusal of a government's money
(f) Action by Governments
92. Domestic embargo
93. Blacklisting of traders
94. International sellers' embargo
95. International buyers' embargo
96. International trade embargo

IV. Economic Noncooperation: Strikes
(a) Symbolic Strikes
97. Protest strike
98. Quickie walkout (lightning strike)
(b) Agricultural Strikes
99. Peasant strike

   100. Farm Workers' strike
 (c) Strikes by Special Groups
   101. Refusal of impressed labor
   102. Prisoners' strike
   103. Craft strike
   104. Professional strike
 (d) Ordinary Industrial Strikes
   105. Establishment strike
   106. Industry strike
   107. Sympathetic strike
 (e) Restricted Strikes
   108. Detailed strike
   109. Bumper strike
   110. Slowdown strike
   111. Working-to-rule strike
   112. Reporting "sick" (sick-in)
   113. Strike by resignation
   114. Limited strike
   115. Selective strike
 (f) Multi-Industry Strikes
   116. Generalized strike
   117. General strike
 (g) Combination of Strikes and Economic Closures
   118. Closing shops or suspending work (hartal)
   119. Economic shutdown
V. Political Noncooperation
 (a) Rejection of Authority
   120. Withholding or withdrawal of allegiance
   121. Refusal of public support
   122. Literature and speeches advocating resistance
 (b) Citizens' Noncooperation with Government
   123. Boycott of legislative bodies
   124. Boycott of elections
   125. Boycott of government employment and positions
   126. Boycott of government departments, agencies, and other bodies
   127. Withdrawal from government educational institutions
   128. Boycott of government-supported organizations
   129. Refusal of assistance to enforcement agents
   130. Removal of own signs and placemarks
   131. Refusal to accept appointed officials
   132. Refusal to dissolve existing institutions
 (c) Citizens' Alternatives to Obedience
   133. Reluctant and slow compliance
   134. Nonobedience in absence of direct supervision
   135. Popular nonobedience

136. Disguised disobedience
137. Refusal of an assemblage or meeting to disperse
138. Sitdown
139. Noncooperation with conscription and deportation
140. Hiding, escape, and false identities
141. Civil disobedience of "illegitimate" laws

(d) Action by Government Personnel
142. Selective refusal of assistance by government aides
143. Blocking of lines of command and information
144. Stalling and obstruction
145. General administrative noncooperation
146. Judicial noncooperation
147. Deliberate inefficiency and selective noncooperation by enforcement agents
148. Mutiny

(e) Domestic Governmental Action
149. Quasi-legal evasions and delays
150. Noncooperation by constituent governmental units

(f) International Governmental Action
151. Changes in diplomatic and other representations
152. Delay and cancellation of diplomatic events
153. Withholding of diplomatic recognition
154. Severance of diplomatic relations
155. Withdrawal from international organizations
156. Refusal of membership in international bodies
157. Expulsion from international organizations

VI. Nonviolent Intervention
(a) Psychological Intervention
158. Self-exposure to the elements
159. The fast (Fast of moral pressure, Hunger strike, Satyagrahic fast)
160. Reverse trial
161. Nonviolent harassment

(b) Physical Intervention
162. Sit-in
163. Stand-in
164. Ride-in
165. Wade-in
166. Mill-in
167. Pray-in
168. Nonviolent raids
169. Nonviolent air raids
170. Nonviolent invasion
171. Nonviolent interjection
172. Nonviolent obstruction
173. Nonviolent occupation

(c) Social Intervention
   174.   Establishing new social patterns
   175.   Overloading of facilities
   176.   Stall-in
   177.   Speak-in
   178.   Guerrilla theater
   179.   Alternative social institutions
   180.   Alternative communication system
(d) Economic Intervention
   181.   Reverse strike
   182.   Stay-in strike
   183.   Nonviolent land seizure
   184.   Defiance of blockades
   185.   Politically motivated counterfeiting
   186.   Preclusive purchasing
   187.   Seizure of assets
   188.   Dumping
   189.   Selective patronage
   190.   Alternative markets
   191.   Alternative transportation systems
   192.   Alternative economic institutions
(e) Political Intervention
   193.   Overloading of administrative systems
   194.   Disclosing identities of secret agents
   195.   Seeking imprisonment
   196.   Civil disobedience of "neutral" laws
   197.   Work-on without collaboration
   198.   Dual sovereignty and parallel government

Gene Sharp's 198 methods of nonviolent action compiled in 1973 remains highly relevant today in spite of the technological advances in the past four decades. In fact, social media and the Internet can help amplify and empower nonviolent action. While some of the 198 methods are quite unconventional, many of them have been practiced by political dissidents around the world. For example, hundreds of thousands of demonstrators employed "47. Assemblies of protest or support" and "18. Displays of flags and symbolic colors" during the 2004 Orange Revolution in Ukraine [73]. In 2011, Egyptian and Arab Spring protesters wrote anti-government songs, posted them on the Internet, and sang them in the streets— just like Gene Sharp suggested in "36. Performances of plays and music" and "37. Singing" [74].

In particular, "181. Reverse strike" under economic intervention is exceptionally constructive. In 1958, Italian social activist Danilo Dolci, known as the Sicilian Gandhi, dispatched 150 unemployed men to repair a dirt road outside Partinico, Sicily when the government had refused to do so or to grant them permission [75]. Dolci was arrested and sentenced to eight months in jail as a result.

Interestingly, "23. Destruction of own property" under symbolic public acts is undeniably destructive, but the key difference between this and terrorism is that it deals with one's own property and not with other people's property or putting others in harm's way. As long as the owner does not endanger anyone or make an insurance claim, destruction of own property can be a powerful selfless statement.

11.10 "We Do Not Have the Right to Resort to Violence When We Don't Get Our Way": President Bill Clinton

The 2011 revolt in Syria began with peaceful protests but it turned into a civil war after the government waged a brutal crackdown on dissent, killing more than 40,000 activists [76]. Armed conflict may be inevitable in some circumstances against an oppressive authoritative regime. However, there is no excuse for violence in a democratic society that welcomes free speech and open debates. In an April 2010 article published in *The New York Times*, President Bill Clinton wrote [77]:

> We should never forget what drove the bombers, and how they justified their actions to themselves. They took to the ultimate extreme an idea advocated in the months and years before the bombing by an increasingly vocal minority: the belief that the greatest threat to American freedom is our government, and that public servants do not protect our freedoms, but abuse them. On that April 19, the second anniversary of the assault of the Branch Davidian compound near Waco, deeply alienated and disconnected Americans decided murder was a blow for liberty.
>
> Americans have more freedom and broader rights than citizens of almost any other nation in the world, including the capacity to criticize their government and their elected officials. But *we do not have the right to resort to violence—or the threat of violence—when we don't get our way*. Our founders constructed a system of government so that reason could prevail over fear. Oklahoma City proved once again that without the law there is no freedom.
>
> Criticism is part of the lifeblood of democracy. No one is right all the time. But we should remember that there is a big difference between criticizing a policy or a politician and demonizing the government that guarantees our freedoms and the public servants who enforce our laws. Civic virtue can include harsh criticism, protest, even civil disobedience. But not violence or its advocacy. That is the bright line that protects our freedom. It has held for a long time, since President George Washington called out 13,000 troops in response to the Whiskey Rebellion. Fifteen years ago, the line was crossed in Oklahoma City. In the current climate, with so many threats against the president, members of Congress and other public servants, we owe it to the victims of Oklahoma City, and those who survived and responded so bravely, not to cross it again.

Oklahoma City bomber Timothy McVeigh never expressed any remorse despite facing execution by lethal injection in May 2001 [78]. However, Centennial Olympic Park bomber Eric Rudolph wrote his mother from jail: "Perhaps I should have found a peaceful outlet for my opposition to the government in Washington: maybe I should have been a lawyer and fought [for] decency in the face of this rotten system; perhaps I could have taken up teaching and sought to inculcate a healthy outlook in a decidedly unhealthy society. But I didn't do any of these things, and I resorted to force to have my voice heard" [79].

Terrorism does not advance the cause but only serves to hurt the intention of its perpetrators. Terrorism should be replaced by nonviolent civil disobedience. Peace journalist Robert Koehler posted in his blog on March 3, 2011: "The 'street' and the 'masses' have actually found—maybe rediscovered is the right word—the power of nonviolent collective action. And the cradle of this new approach to civilization is not in the comfortably snoozing West but in the impoverished and long-brutalized Middle East, where every despot has either tumbled or is shaking in his boots—and where violence has suddenly been stripped of its righteousness and been exposed as weakness, no matter how much mayhem it produces" [80].

When terrorists realize that there are many more negatives in conducting terrorism and many more positives in abandoning terrorism, the horrible disease will be cured. In an unauthenticated document dated December 14, 2001 (three months after the 9/11 attacks), Osama bin Laden predicted in his will that he would be killed as a result of a "betrayal" and he instructed his children not to continue to wage the holy war against America and Israel [81].

11.11 "Peace is the only path to true security": President Barack Obama

Five days after the Sandy Hook Elementary School massacre in Newtown, Connecticut, President Barack Obama spoke in a White House press conference on December 19, 2012: "It's encouraging that people of all different backgrounds and beliefs and political persuasions have been willing to challenge some old assumptions and change some long-standing positions. That conversation has to continue, but this time the words need to lead to action" [82]. Obama's words apply not only to finding long-term solutions to gun violence in the U.S. but also to domestic and international terrorism around the globe.

During his visit to Israel and the West Bank in March 2013, President Obama told a captive audience at the International Convention Center in Jerusalem [83]:

"Peace is necessary. Peace is the only path to true security. ... There is no question that Israel has faced Palestinian factions who turned to terror, and leaders who missed historic opportunities. That is why security must be at the center of any agreement. And there is no question that the only path to peace is through negotiation. ... The only way to truly protect the Israeli people over the long term is through the absence of war - because no wall is high enough, and no Iron Dome is strong enough or perfect enough, to stop every enemy from inflicting harm. ... Peace must be made among peoples, not just governments. No single step can change overnight what lies in the hearts and minds of millions. No single step is going to erase years of history and propaganda. But progress with the Palestinians is a powerful way to begin, while sidelining extremists who thrive on conflict and division. It would make a difference!"

Indeed, it would make a positive difference in world security and counterterrorism by setting our mind on pursuing peaceful solutions rather than escalating the war on terror.

References

1. **CDC.** Parasites - Dracunculiasis (also known as Guinea Worm Disease). [Online] Centers for Disease Control and Prevention, November 21, 2012. http://www.cdc.gov/parasites/guineaworm/.
2. **The Carter Center.** Guinea Worm Disease Eradication. [Online] The Carter Center, 2011. http://www.cartercenter.org/health/guinea_worm/mini_site/index.html.
3. **Dougherty, Jill.** Experts: No easy cure for the disease of terror. [Online] CNN, July 27, 2012. http://security.blogs.cnn.com/2012/07/27/experts-no-easy-cure-for-the-disease-of-terror/.
4. **CBS News.** Ex-CIA Operative Comes Out of the Shadows. [Online] CBS News, August 2, 2010. http://www.cbsnews.com/8301-18560_162-6014887.html.
5. **Thoreau, Henry David.** Walden; or, Life in the Woods. [Online] Ticknor and Fields, August 9, 1854.
6. **TED.** The technological future of crime: Marc Goodman at TEDGlobal 2012. [Online] TED, June 28, 2012. http://blog.ted.com/2012/06/28/the-technological-future-of-crime-marc-goodman-at-tedglobal-2012/.
7. **Hammurabi.** Hammurabi's Code: An Eye for an Eye. [Online] ushistory.org, 1792-50 B.C. http://www.ushistory.org/civ/4c.asp.
8. **Orwell, George.** Revenge is Sour. [Online] The Tribune, November 9, 1945. http://www.george-orwell.org/Revenge_is_Sour/0.html.
9. **Robinson, Adam.** Bin Laden: Behind the Mask of a Terrorist. [Online] Arcade Publishing, November 22, 2002. http://books.google.com/books?id=zTGPHuW4qGIC&pg=PT118.
10. **Chief of Naval Operations.** The United States Navy In "Desert Shield"/"Desert Storm". [Online] Naval History & Heritage, May 15, 1991. http://www.history.navy.mil/wars/dstorm/index.html.
11. **Schmitt, Eric.** U.S. to Withdraw All Combat Forces From Saudi Arabia. [Online] The New York Times, April 29, 2003. http://www.nytimes.com/2003/04/29/international/worldspecial/29CND-RUMS.html.
12. **Ray, Julie.** Opinion Briefing: U.S. Image in Middle East/North Africa. [Online] Gallup, January 27, 2009. http://www.gallup.com/poll/114007/opinion-briefing-image-middle-east-north-africa.aspx.
13. **Starr, Barbara.** How the Patriot deployment to Turkey will work. [Online] CNN, December 14, 2012. http://security.blogs.cnn.com/2012/12/14/how-the-patriot-deployment-to-turkey-will-work/.
14. **Quilliam.** The Attack on the US Consulate Was A Planned Terrorist Assault Against US and Libyan Interests. [Online] Quilliam, September 12, 2012. http://www.quilliamfoundation.org/press-releases/the-attack-on-the-us-consulate-was-a-planned-terrorist-assault-against-us-and-libyan-interests/.
15. **Zway, Suliman Ali and Fahim, Kareem.** Angry Libyans Target Militias, Forcing Flight. [Online] The New York Times, September 21, 2012. http://www.nytimes.com/2012/09/22/world/africa/pro-american-libyans-besiege-militant-group-in-benghazi.html.
16. **Bacon, Francis Sir.** Essayes and counsels, civil and moral. [Online] 1664. http://www.folger.edu/eduPrimSrcDtl.cfm?psid=123.
17. **The Metropolitan Museum of Art.** The Crusades (1095–1291). [Online] The Metropolitan Museum of Art. [Cited: December 9, 2012.] http://www.metmuseum.org/toah/hd/crus/hd_crus.htm.
18. **IMDb.** Memorable quotes for Star Trek II: The Wrath of Khan. [Online] IMDb, 1982. http://www.imdb.com/title/tt0084726/quotes.
19. **Ackman, Dan.** The Cost Of Being Osama Bin Laden. [Online] Forbes, September 14, 2001. http://www.forbes.com/2001/09/14/0914ladenmoney.html.
20. **The Economist.** Osama bin Laden. [Online] The Economist, May 5, 2011. http://www.economist.com/node/18648254.
21. **Corera, Gordon.** MI5 head warns of serious risk of UK terrorist attack. [Online] BBC News, September 16, 2010. http://www.bbc.co.uk/news/uk-11335412.
22. **Robertson, Nic and Cruickshank, Paul.** Cagefighter 'cures' terrorists. [Online] CNN, July 23, 2012. http://www.cnn.com/2012/07/20/world/europe/uk-caging-terror-main/index.html.

23. **Robertson, Nic and Cruickshank, Paul.** Convicted terrorist calmed by cagefighting. [Online] CNN, July 28, 2012. http://edition.cnn.com/2012/07/22/world/europe/uk-caging-terror-mansha/index.html.

24. **Ahmed, Saeed, Cohen, Lisa and Sidner, Sara.** From horror to hope: Boy's miracle recovery from brutal attack. [Online] CNN, December 7, 2012. http://www.cnn.com/2012/12/06/world/freedom-project-operation-hope/index.html.

25. **Nelson, Dean.** Slumdog Millionaire: Meet the real Mumbai street urchins. [Online] The Telegraph, January 18, 2009. http://www.telegraph.co.uk/news/worldnews/asia/india/4280812/Slumdog-Millionaire-Meet-the-real-Mumbai-street-urchins.html.

26. **msnbc.com.** WikiLeaks: U.K. trained Bangladeshi 'death squad'. [Online] MSNBC, December 21, 2010. http://www.msnbc.msn.com/id/40773855/ns/us_news-wikileaks_in_security/.

27. **Ghosh, Bobby.** After Waterboarding: How to Make Terrorists Talk? [Online] Time Magazine, June 8, 2009. http://www.time.com/time/magazine/article/0,9171,1901491,00.html.

28. **Wright, Lawrence.** The Agent. Did the C.I.A. stop an F.B.I. detective from preventing 9/11? [Online] The New Yorker, July 10, 2006. http://www.newyorker.com/archive/2006/07/10/060710fa_fact_wright?currentPage=all.

29. **60 Minutes.** Ex-FBI agent who interrogated Qaeda members speaks out. [Online] CBS Interactive Inc, September 9, 2011. http://www.cbsnews.com/8301-18560_162-20104007/ex-fbi-agent-who-interrogated-qaeda-members-speaks-out.

30. **Stavridis, James.** James Stavridis: A Navy Admiral's thoughts on global security. [Online] TED, June 2012. http://www.ted.com/talks/james_stavridis_how_nato_s_supreme_commander_thinks_about_global_security.html.

31. **Dear, John S.J.** Afghanistan journal, part one: Learning a nonviolent lifestyle in Kabul. [Online] National Catholic Reporter, December 11, 2012. http://ncronline.org/blogs/road-peace/afghanistan-journal-part-one-learning-nonviolent-lifestyle-kabul.

32. **Dear, John S.J.** Afghanistan journal, part two: bearing witness to peacemaking in a war-torn country. [Online] National Catholic Reporter, December 18, 2012. http://ncronline.org/blogs/road-peace/afghanistan-journal-part-two-bearing-witness-peacemaking-war-torn-country.

33. **Adichie, Chimamanda.** Chimamanda Adichie: The danger of a single story. [Online] TED, October 2009. http://www.ted.com/talks/chimamanda_adichie_the_danger_of_a_single_story.html.

34. **Engber, Daniel.** I'm Covered in Leaflets! [Online] Slate, July 18, 2006. http://www.slate.com/articles/news_and_politics/explainer/2006/07/im_covered_in_leaflets.html.

35. **Digital Daya.** Research Note: World Leaders on Twitter. [Online] Digital Daya, December 2012. http://www.digitaldaya.com/admin/modulos/galeria/pdfs/69/156_biqz7730.pdf.

36. **Press Association.** David Cameron gets 100,000 Twitter followers days after starting account. [Online] The Guardian, October 9, 2012. http://www.guardian.co.uk/politics/2012/oct/09/david-cameron-100000-twitter-followers.

37. **Atkins, Christopher R.** "I solemnly swear to defend the Constitution of the United States of America against all enemies, foreign and domestic.". [Online] Michael Moore, January 28, 2005. http://www.michaelmoore.com/words/soldiers-letters/i-solemnly-swear-to-defend-the-constitution-of-the-united-states-of-america-against-all-enemies-foreign-and-domestic.

38. **Kansas high school students.** We Are Hungry. [Online] YouTube, September 17, 2012. http://www.youtube.com/watch?v=2IB7NDUSBOo.

39. **Cohen, Elizabeth.** Peanut butter, garlic bread back on school plates. [Online] CNN, December 12, 2012. http://www.cnn.com/2012/12/12/health/school-lunch-changes/index.html.

40. **Facebook.** Form S-1 Registration Statement. [Online] United States Securities and Exchange Commission, February 1, 2012. http://sec.gov/Archives/edgar/data/1326801/000119312512034517/d287954ds1.htm.

41. **Levine, Adam.** A social network site for jihadists? [Online] CNN, April 5, 2002. http://security.blogs.cnn.com/2012/04/05/faqebook-dreams-of-a-jihadi-social-network/.

42. **Drash, Wayne.** Bringing healing to Newtown, one pie at a time. [Online] CNN, December 19, 2012. http://eatocracy.cnn.com/2012/12/19/bringing-healing-to-newtown-one-pie-at-a-time/.

43. **Frontline.** News War. [Online] wgbh educational foundation. [Cited: December 11, 2012.] http://www.pbs.org/wgbh/pages/frontline/teach/newswar/hand1.html.

44. **Bohlen, Celestine.** In New War on Terrorism, Words Are Weapons, Too. [Online] The New York Times, September 29, 2001. http://www.nytimes.com/2001/09/29/arts/think-tank-in-new-war-on-terrorism-words-are-weapons-too.html.

45. **ABC News.** Maher Apologizes for 'Cowards' Remark. [Online] ABC News, September 20, 2001. http://abcnews.go.com/Entertainment/story?id=102318&page=1.

46. **Apple, R. W. Jr.** Pentagon Papers. [Online] The New York Times, June 23, 1996. http://topics.nytimes.com/top/reference/timestopics/subjects/p/pentagon_papers/index.html.

47. **Haberman, Clyde.** Arthur O. Sulzberger, Publisher Who Transformed The Times for New Era, Dies at 86. [Online] 2012, 29 September. http://www.nytimes.com/2012/09/30/nyregion/arthur-o-sulzberger-publisher-who-transformed-times-dies-at-86.html?pagewanted=all.

48. **Peters, Jeremy W.** Latest Word on the Trail? I Take It Back. [Online] The New York Times, July 15, 2012. http://www.nytimes.com/2012/07/16/us/politics/latest-word-on-the-campaign-trail-i-take-it-back.html?pagewanted=all.

49. **Rather, Dan.** Dan Rather: 'Quote approval' a media sellout. [Online] CNN, July 19, 2012. http://www.cnn.com/2012/07/19/opinion/rather-quote-approval-reporting/index.html.

50. **IMDb.** Fahrenheit 9/11. [Online] IMDb, June 25, 2004. http://www.imdb.com/title/tt0361596/.

51. **Rutenberg, Jim.** Disney Is Blocking Distribution of Film That Criticizes Bush. [Online] The New York Times, May 5, 2004. http://www.nytimes.com/2004/05/05/us/disney-is-blocking-distribution-of-film-that-criticizes-bush.html.

52. **Keller, Bill.** WikiLeaks, a Postscript. [Online] The New York Times, February 19, 2012. http://www.nytimes.com/2012/02/20/opinion/keller-wikileaks-a-postscript.html.

53. **Ellsberg, Daniel.** Daniel Ellsberg on Colbert Report: Julian Assange is Not a Criminal Under the Laws of the United States. [Online] Ellsberg.Net, December 10, 2010. http://www.ellsberg.net/archive/daniel-ellsberg-on-colbert-report.

54. **Blanton, Thomas.** Hearing on the Espionage Act and the Legal and Constitutional Implications of Wikileaks. [Online] Committee on the Judiciary, U.S. House of Representatives, December 16, 2010. http://www.gwu.edu/~nsarchiv/news/20101216/Blanton101216.pdf.

55. **RT.** WikiLeaks revelations only tip of iceberg – Assange. [Online] RT, May 3, 2011. http://rt.com/news/wikileaks-revelations-assange-interview/.

56. **Bulwer-Lytton, Edward.** Richelieu, or The conspiracy: in five acts. [Online] Chapman and Hall, 1856. http://books.google.com/books?id=FfktAAAAYAAJ.

57. **Holmes, Oliver Wendell Jr.** Schenck v. United States. [Online] Cornell University Law School, March 3, 1919. http://www.law.cornell.edu/supct/html/historics/USSC_CR_0249_0047_ZO.html.

58. **FBI.** 1919 Bombings. [Online] Federal Bureau of Investigation. [Cited: December 12, 2012.] http://www.fbi.gov/philadelphia/about-us/history/famous-cases/famous-cases-1919-bombings.

59. **Supreme Court of the United States.** Brandenburg v. Ohio. [Online] Cornell University Law School, June 9, 1969. http://www.law.cornell.edu/supct/html/historics/USSC_CR_0395_0444_ZO.html.

60. **CNN Wire Staff.** No let-up in protests over anti-Islam film. [Online] CNN, September 18, 2012. http://www.cnn.com/2012/09/17/world/film-protests/index.html.

61. U.S. Missions Stormed in Libya, Egypt. [Online] Bradley, Matt; Nissenbaum, Dion, September 12, 2012. http://online.wsj.com/article/SB10000872396390444017504577645681057498266.html.

62. **York, Jillian C.** Should Google censor an anti-Islam video?. [Online] CNN, September 16, 2012. http://www.cnn.com/2012/09/14/opinion/york-libya-youtube/index.html.

63. **Madhani, Aamer.** Cleric al-Awlaki dubbed 'bin Laden of the Internet'. [Online] USA Today, August 24, 2010. http://usatoday30.usatoday.com/news/nation/2010-08-25-1A_Awlaki25_CV_N.htm.

64. **Gardham, Duncan, Rayner, Gordon and Bingham, John.** Why hasn't YouTube taken down terror videos? [Online] The Telegraph, November 3, 2010. http://www.telegraph.co.uk/news/politics/8106672/Why-hasnt-YouTube-taken-down-terror-videos.html.

65. **Basu, Moni.** New details emerge of anti-Islam film's mystery producer. [Online] CNN, September 14, 2012. http://www.cnn.com/2012/09/13/world/anti-islam-filmmaker/index.html.

66. **Cawthon, Erinn.** Controversial 'Defeat Jihad' ad to appear in NYC subways. [Online] CNN, September 19, 2012. http://www.cnn.com/2012/09/19/us/new-york-controversial-subway-ad/index.html.

67. **CBS Radio New York.** Bloomberg Weighs In On Provocative Subway Ad. [Online] CBS Radio New York, September 21, 2012. http://newyork.cbslocal.com/2012/09/21/bloomb erg-weighs-in-on-provocative-subway-ad/.
68. **Kahn-Troster, Rachel.** Subway ads: A right to hate speech, a duty to condemn. [Online] CNN, September 25, 2012. http://www.cnn.com/2012/09/25/opinion/kahn-troster-anti-islam-hate-ads/ index.html.
69. **Gandhi, Mohandas K.** An Autobiography: The Story of My Experiments with Truth. [Online] Beacon Press, 1993. http://books.google.com/books/about/An_Autobiography.html ?id=VsMLYjEsyaEC.
70. **Sgueglia, Kristina.** Interfaith group protests ad that says 'Support Israel. Defeat Jihad'. [Online] CNN, September 25, 2012. http://religion.blogs.cnn.com/2012/09/25/ interfaith-group-protests-ad-that-says-support-israel-defeat-jihad/.
71. **Sharp, Gene.** 198 Methods of Nonviolent Action. [Online] Porter Sargent Publishers, 1973. http://www.aeinstein.org/organizations103a.html.
72. **Sharp, Gene.** 198 Methods of Nonviolent Action. [Online] The Albert Einstein Institution. [Cited: December 27, 2012.] http://www.aeinstein.org/organizations/org/198_methods.pdf.
73. **Meek, James.** Divided they stand. [Online] The Guardian, December 9, 2004. http://www. guardian.co.uk/world/2004/dec/10/ukraine.jamesmeek.
74. **Lee, Amy.** Egypt's Revolutionary Music, And 7 Other Revolutions That Turned To Song. [Online] The Huffington Post, January 25, 2012. http://www.huffingtonpost.com/2012/01/25/ egypt-revolution-january-25_n_1229332.html.
75. **Tagliabue, John.** Danilo Dolci, Vivid Voice Of Sicily's Poor, Dies at 73. [Online] The New York Times, December 31, 1997. http://www.nytimes.com/1997/12/31/ world/danilo-dolci-vivid-voice-of-sicily-s-poor-dies-at-73.html.
76. **The Associated Press.** Syria shuts down Internet access as country imposes nationwide online blackout. [Online] New York Daily News, November 29, 2012. http://www.nydailynews.com/ news/world/nationwide-internet-blackout-syria-article-1.1210074.
77. **Clinton, Bill.** What We Learned in Oklahoma City. [Online] The New York Times, April 18, 2010. http://www.nytimes.com/2010/04/19/opinion/19clinton.html.
78. **Borger, Julian.** McVeigh brushes aside deaths. [Online] The Guardian, March 29, 2001. http://www.guardian.co.uk/world/2001/mar/30/julianborger.
79. **Morrison, Blake.** Special report: Eric Rudolph writes home. [Online] USA Today, July 5, 2005. http://usatoday30.usatoday.com/news/nation/2005-07-05-rudolph-cover-partone_x.htm.
80. **Koehler, Robert.** You Can't Kill an Idea. [Online] The Huffington Post, March 3, 2011. http://www.huffingtonpost.com/robert-koehler/you-cant-kill-an-idea_b_830881.html.
81. **Flock, Elizabeth.** Osama bin Laden tells his children not to fight jihad in his will. [Online] The Washington Post, May 4, 2011. http://www.washingtonpost.com/blogs/blogpost/ post/osama-bin-laden-tells-children-not-to-fight-jihad-in-his-will/2011/05/04/AFDP4UmF_b log.html.
82. **Obama, Barack.** Remarks by the President in a Press Conference. [Online] The White House, December 19, 2012. http://www.whitehouse.gov/the-press-office/2012/12/19/ remarks-president-press-conference.
83. **Miller, Sara.** Obama: Peace is the only path to true security. [Online] The Jerusalem Post, March 21, 2013. http://www.jpost.com/Diplomacy-and-Politics/Obama-Peace-is-the-only-path-to-true-security-307323

Chapter 12
War and Peace

Violence never brings permanent peace. It solves no social problem: it merely creates new and more complicated ones.
—Martin Luther King Jr.
Nobel Peace Prize acceptance speech (December 11, 1964).

Too many of us think [Peace] is impossible. Too many think it unreal. But that is a dangerous, defeatist belief. ... Our problems are manmade—therefore they can be solved by man [or woman].
—President John F. Kennedy (June 10, 1963).

The belief that peace is desirable is rarely enough to achieve it. Peace requires responsibility. Peace entails sacrifice.
—President Barack Obama.
Nobel Peace Prize acceptance speech (December 10, 2009).

If there is an Internet connection, my camera is more powerful [than my AK-47].
—Syrian dissident Abu Ghassan (June 2012).

Instead of building walls to create security, we need to build bridges.
—U.S. Navy Admiral and NATO's Supreme Allied Commander James Stavridis, TEDGlobal 2012 (June 2012).

12.1 War as State-Sponsored Terrorism

The Federal Bureau of Investigation (FBI) defines terrorism as "the unlawful use of force or violence against persons or property to intimidate or coerce a government, the civilian population, or any segment thereof in furtherance of political or social objectives" [1]. MIT Professor Emeritus Noam Chomsky believes that

N. Lee, *Counterterrorism and Cybersecurity*, DOI: 10.1007/978-1-4614-7205-6_12,   197
© Springer Science+Business Media New York 2013

the U.S. official doctrine of low-intensity warfare is almost identical to the official definition of terrorism [2]. Political commentator Bill Maher equates U.S. drone attacks with terrorist acts [3].

While a terrorist act causes innocent people pain, suffering, and even death, war is legitimized state-sponsored terrorism in a grand scale. In World War II, over 50 million people lost their lives, more than half of whom were civilians who perished in bombings, by famine, and other war-related circumstances [4]. Between five and six million Jews were killed in the Holocaust [5]. Over 27 % of the civilian population in Hiroshima and 24 % of the residents in Nagasaki were wiped out by atomic bombs [6].

In war-torn countries, people live in constant fear. Jesuit priest and peace activist John Dear interviewed families at the refugee camps in Afghanistan in December 2012. Raz Mohammad, a member of Afghan Peace Volunteers, told his somber story [7]:

> My brother-in-law was killed by a U.S. drone in 2008. He was a student, and visiting some friends one summer evening when they decided to walk to a garden and sit there and talk. They were enjoying the evening, sitting in the garden, when a drone flew by and dropped a bomb. Everyone was incinerated. We couldn't find any remains. My sister was left behind with her baby boy. I think the drone attacks were first begun in my province. We hear them about every three nights. They have a low, buzzing sound, like a mosquito. They hover over us. They fly over us during the day, and fly over us during the night, when we can see the spotlight at the front of the drone. Occasionally, the large U.S. fighter bombers fly over, and they make a huge noise. All the people of the area, especially the children, are afraid of the U.S. soldiers, the U.S. tanks, the U.S. drones, and the U.S. fighter bombers. They fear being killed.
>
> ...
>
> No one I know wants the war to continue. Ordinary people everywhere are sick and tired of war, yet we're demonized as warriors and terrorists. None of us can tell who is a member of the Taliban and who isn't. If we can't tell who is a member of the Taliban, how can anyone in the U.S. claim to know who is in the Taliban? Meanwhile, our schools, hospitals and local services have all collapsed. The U.S./NATO forces are not helping anyone, only bringing fear and death to the people.

At a women's sewing cooperative in Afghanistan, a woman expressed her frustration and pleaded: "I thought President Obama would care for the oppressed, but he has made things much worse for us. He is even worse than President Bush. Please ask the people of the U.S. to take to the streets again and do what they can to stop this war now" [7].

The woman's plea did not fall on deaf ears. In January 2013, White House officials said that the Obama administration is considering the withdrawal of all U.S. troops from Afghanistan after Operation Enduring Freedom in Afghanistan (OEF-A) officially ends in late 2014 [8].

In a way, President Barack Obama acknowledged his predicament when he accepted the 2009 Nobel Peace Prize amid controversy and in the midst of two wars. Obama said, "We are at war, and I'm responsible for the deployment of thousands of young Americans to battle in a distant land. Some will kill, and some will be killed. And so I come here with an acute sense of the costs of armed conflict—filled with difficult questions about the relationship between war and peace, and our effort to replace one with the other. ... To say that force may sometimes be necessary is not a call to cynicism—it is a recognition of history; the imperfections of man and the limits of reason" [9].

12.2 Complacency in War

At the Battle of Fredericksburg on December 13, 1862, U.S. Confederate General Robert E. Lee said to his men, "It is well that war is so terrible, otherwise we should grow too fond of it" [10]. Notwithstanding the horrors of war, desensitization to violence has perpetuated armed conflicts around the world, and the public has become inured to all but the most catastrophic or apocalyptic events.

Take the decades-long Gaza-Israel conflict as an example. Both Israelis and Palestinians have died from rocket attacks, airstrikes, and shootings. When the conflict escalated and death toll rose in November 2012, U.S. Secretary of State Hillary Clinton went to Jerusalem to meet with Israel Prime Minister Benjamin Netanyahu to push for a truce [11]. When a ceasefire was announced in Cairo, both sides claimed victory and accused the other as the oppressor. Israel killed Hamas' military leader Ahmed al-Jaabari and significantly weakened their offensive capability [12], whereas Hamas declared triumph with diplomatic support from Egypt, Turkey, and Qatar [13].

In his Nobel Peace Prize acceptance speech on December 11, 1964, Martin Luther King Jr. said, "Nations have frequently won their independence in battle. But in spite of temporary victories, violence never brings permanent peace. It solves no social problem: it merely creates new and more complicated ones. Violence is impractical because it is a descending spiral ending in destruction for all. It is immoral because it seeks to humiliate the opponent rather than win his understanding: it seeks to annihilate rather than convert" [14].

Win or lose, innocent civilians are caught in the middle of the deadly violence [15]. "Innocent people, including children, have been killed or injured on both sides. Families on both sides were forced to cower in fear as the violence raged around them," said U.N. Secretary-General Ban Ki-moon [16].

In an armed conflict, the International Humanitarian Law protects "those who do not take part in the fighting, such as civilians and medical and religious military personnel" [17]. A major part of the law is contained in the four Geneva Conventions of 1949 and nearly every country in the world has agreed to be bound by them. With that in mind, the United Nations fact-finding mission headed by Judge Richard Goldstone published a 574-page report in September 2009 accusing both Hamas and Israel of deliberately targeting civilians [18]. Goldstone later retracted the findings, but a senior Israel military official described in November 2012 that civilian casualties as "regrettable" but unavoidable because the "terrorist infrastructure is embedded inside the population" [19].

Politicians and the military-industrial complex have grown to be complacent about war. Violence has become the first course of action and the first choice of reaction in dealing with international conflicts, even at the expense of innocent civilians. In fact, according to American and Israeli officials, the November 2012 Gaza-Israel conflict was considered "a practice run for any future armed confrontation with Iran, featuring improved rockets that can reach Jerusalem and new antimissile systems to counter them" [20].

War has become a popular option in the political playbook, a child's play with deadly consequences. An elder in an Afghanistan refugee camp put it this way, "The powers that be have turned Afghanistan into a killing field, their personal playground of war" [7].

In his 1869 novel *War and Peace*, Leo Tolstoy warned that "Война не любезность, а самое гадкое дело в жизни, и надо понимать это и не играть в войну. (War is not a courtesy but the most horrible thing in life; and we ought to understand that, and not play at war.)" [21].

12.3 Civilians Wanting Peace

In the 2011 film *Sherlock Holmes: A Game of Shadows*, Professor James Moriarty intoned, "You see, hidden within the unconscious is an insatiable desire for conflict. So you're not fighting me so much as you are the human condition. All I want is to supply the bullets and the bandages. War on an industrial scale is inevitable" [22].

Perhaps Professor Moriarty strikes a chord with the military-industrial complex, but civilians want peace more than anything else. Afghan Peace Volunteer Raz Mohammad told Jesuit priest John Dear, "We should not accept these drone attacks if we are human beings. They are killing innocent human beings. Humanity should not allow this to happen. No one I know wants the war to continue. Ordinary people everywhere are sick and tired of war" [7].

Doctors at the Sheba Medical Center at a Tel Aviv hospital treated a four-year-old boy from Israel and an eight-year-old girl from Gaza who both lost fingers from rocket blasts. Dr. Batia Yaffe remarked, "I come to think about what is it about this piece of land that everybody is fighting about it all the time. This is what comes to my mind: whether this is our lot for eternity from now on. Always have injuries on both sides, always fighting—what's the point?" [23].

12.4 Peace Entailing Sacrifice

Sherlock Holmes replied to Professor James Moriarty, "A winning strategy sometimes necessitates sacrifice. A war has been averted" [22].

In his Nobel Peace Prize acceptance speech on December 10, 2009, President Barack Obama said, "In many countries, there is a disconnect between the efforts of those who serve and the ambivalence of the broader public. I understand why war is not popular, but I also know this: The belief that peace is desirable is rarely enough to achieve it. Peace requires responsibility. Peace entails sacrifice" [9].

Being the President of the United States is itself a sacrifice. The job is highly competitive and extremely stressful, and yet it does not pay as well as most CEO's of the Fortune 500 companies. The President enjoys very little privacy in his personal life. Taking a leisure stroll outside the White House or dining in a restaurant requires Secret Service protection during the time of presidency as well as for a period of 10 years from the date the president leaves office [24].

Serving in the military is another sacrifice. Many soldiers are deployed in foreign lands under harsh and life-threatening environments, far away from their friends and families. Many veterans suffer from Post-Traumatic Stress Disorder (PTSD). In 2011, there were 165 active-duty suicides in the U.S. Army [25]. In 2012, the Army reported a record high of 325 suicides among active and non-active military personnel, exceeding the total number of U.S. military combat casualties—313—in Operation Enduring Freedom in Afghanistan [26]. "Suicide is the toughest enemy I have faced in my 37 years in the Army," remarked Gen. Lloyd J. Austin III, vice chief of staff of the Army, in a 2012 news release from the U.S. Department of Defense [27].

We all want peace, but wanting alone is not enough to actuate peace. After the 9/11 terrorist attacks, Disney Online General Manager Ken Goldstein said at a senior staff meeting, "Before 9/11, no one in this room knew what al-Qaeda was." The room was filled with highly-educated and talented senior executives, producers, and engineers at The Walt Disney Company. Busy with work and going about our everyday lives, we did not make time to look around the world. We rely too much on our government to make the right decisions and law enforcement to protect us. Had we been more vigilant, inquisitive, attentive to world affairs, and proactive in politics, the 9/11 attacks might have been averted. As global security advisor and futurist Marc Goodman said at the TEDGlobal 2012 conference, "Public safety is too important to leave to the professionals" [28].

In his 1961 inaugural address, President John F. Kennedy said, "And so, my fellow Americans: ask not what your country can do for you—ask what you can do for your country" [29]. Your country is you, your family, friends, neighbors, and coworkers. The first three words in the Preamble to the United States Constitution are "We the People"—not "We the Government" [30] (See Fig. 12.1).

Kennedy continued, "My fellow citizens of the world: ask not what America will do for you, but what together we can do for the freedom of man. Whether you are citizens of America or citizens of the world, ask of us the same high standards of strength and sacrifice which we ask of you" [29].

Although not every citizen can be as brave as Pakistani schoolgirl Malala Yousafzai who became an activist for education and women's rights at the tender age of 11, everyone should be inspired by her courage and perseverance [31]. In October 2012, the 14 year-old Yousafzai was shot in the head and neck by Taliban gunmen who boarded her school bus [32]. The abhorrent assassination attempt prompted the United Nations to launch the online petition "I Am Malala" to honor Yousufzai and call for Pakistan and countries worldwide to ensure that all children have access to education [33].

U.S. First Lady Laura Bush wrote, "Malala is the same age as another writer, a diarist, who inspired many around the world. From her hiding place in

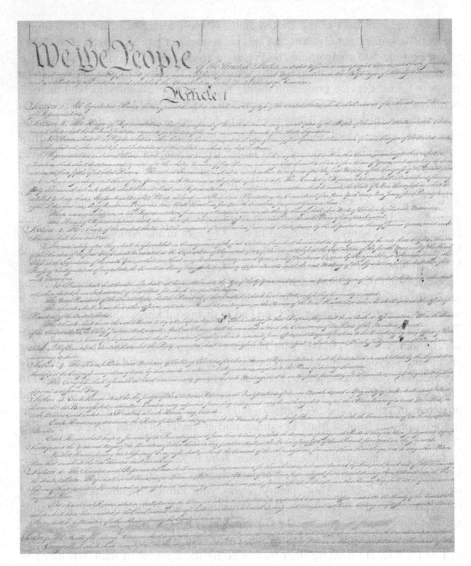

Fig. 12.1   United States Constitution (Page 1 of 4)

Amsterdam, Anne Frank wrote, 'How wonderful it is that nobody need wait a single moment before starting to improve the world.' Today, for Malala and the many girls like her, we need not and cannot wait. We must improve their world" [34].

We must also improve U.S. education. When I was in college, I once played a question card game with my friends. I chose a card in random and the question was: What would be your first order of business if you were elected President of the United States? I gave a succinct answer: Improve the educational system. One of my American-born friends rebuked me immediately, saying that the U.S.

education was perfectly fine as is. A few years later in 1987, University of Chicago Professor Allan Bloom published the book *The Closing of the American Mind* in which he described how "higher education has failed democracy and impoverished the souls of today's students" [35]. Universities are good at producing engineers, doctors, lawyers, and such, but they often gloss over difficult moral and philosophical subjects such as the meaning of life, love, prejudice, war, and peace.

MIT Professor Emeritus Noam Chomsky offered this advice: "You have to try and develop a critical, open mind, and you have to be willing to evaluate and challenge conventional beliefs—accept them if they turn out to be valid but reject them if, as is so often the case, they turn out to just reflect power structures. And then you proceed with educational and organizing activities, actions as appropriate to circumstances. There is no simple formula; rather, lots of options. And gradually over time, things improve. I mean, even the hardest rock will be eroded by steady drips of water. That's what social change comes to and there are no mysterious modes of proceeding. They're hard ones, demanding ones, challenging, often costly. But that's what it takes to get a better world" [36].

12.5 Peace and Friendships on Facebook and Social Media

Peace on Facebook https://peace.facebook.com/ states that "Facebook is proud to play a part in promoting peace by building technology that helps people better understand each other. By enabling people from diverse backgrounds to easily connect and share their ideas, we can decrease world conflict in the short and long term" [37].

According to a Pew Internet report in March 2012, 18 % of social networking site users have blocked, unfriended, or hidden someone on the site because of their differences in political views [38]. On the other hand, Friendships on Facebook publishes daily "friending" numbers between people of different regions, religions, and political affiliations.

Figure 12.2 shows that new friendships are formed everyday regardless of the prevailing political climate. New geographic connections are on the upswing from July 2012 to January 2013 across the board, with India-Pakistan leading the chart by an increase of 120 %. Religious and political connections, however, are down overall by an average of 20 %.

One can observe a 27 % drop in Facebook daily Muslim-Jewish friendships in the aftermath of the Israel-Hamas conflict in November 2012 (See Fig. 12.3). Similarly, the Facebook daily U.S. Conservative/Liberal friendships also decreased by 27 % after the 2012 U.S. Presidential election (See Fig. 12.4).

In June 2012, *Time Magazine*'s congressional correspondent Jay Newton-Small asked the Syrian dissident Abu Ghassan whether his AK-47 or his video camera was the more powerful weapon. Ghassan replied, "My AK!" But he paused for a few seconds, and said, "Actually, if there is an Internet connection, my camera is more powerful" [39]. Partially aided by the Internet Freedom Grants from the U.S.

Fig. 12.2 Facebook daily friendships statistics on July 6, 2012 and January 24, 2013

Geographic Friendships	Connections o July 6, 201	Connections on January 24, 2013
India-Pakistan	168,999	371,790
Albania-Serbia	21,804	38,572
Israel-Palestine	18,321	24,044
Greece-Turkey	5,970	9,695

Religious Friendships	Connections on July 6, 2012	Connections on January 24, 2013
Muslim-Christian	120,542	107,741
Christian-Atheist	32,381	25,638
Muslim-Jewish	545	501
Sunni-Shiite	164	103

Political Friendship	Connections on July 6, 2012	Connections on January 24, 2013
U.S. Conservative/Liberal	8,034	6,064

Fig. 12.3 Facebook daily Muslim-Jewish friendships the day before and after the Israel-Hamas conflict in November 2012

Religious Friendships	Connections on November 13	Connections on November 22
Muslim-Jewish	644	470

Fig. 12.4 Facebook daily U.S. conservative/liberal friendships two days before and after the U.S. Presidential election in November 2012

Political Friendship	Connections on November 4	Connections on November 8
U.S. Conservative/Liberal	9,148	6,716

State Department, Syrian rebels have been filming the protests and posting them on the Internet [40]. Newton-Small called Ghassan a "cyber warrior" [41].

Indeed, the power of the Internet to change the world cannot be understated. The Internet has empowered individuals to start their own grassroots movements. Amid the longstanding Arab–Israeli conflict since 1920 that has taken more than 107,000 lives [42] and has cost more than $12 trillion dollars [43], Ronny Edry, an Israeli graphic designer based in Tel Aviv, reached out to the people of Iran on Facebook in March 2012. Edry and his wife uploaded posters on the Facebook page of Pushpin Mehina with the resounding message "IRANIANS. We will never bomb your country. We *Heart* You" [44].

Edry shared his Facebook experience: "My idea was simple. I was trying to reach the other side. There are all these talks about war, Iran is coming to bomb us and we bomb them back, we are sitting and waiting. I wanted to say the simple words that this war is crazy. In a few hours, I had hundreds of shares and thousands of likes. ... I think it's really amazing that someone from Iran poked me and said 'Hello, I'm from Iran, I saw your poster on Facebook.' ... I got a private message from Iran: 'We love you too. Your word reaches out there, despite the censorship. And Iranian people, aside from the regime, have no hard feelings or animosity towards anybody, particularly Israelis'" [45].

Time Magazine named Facebook co-founder and CEO Mark Zuckerberg its Person of the Year 2010 for connecting more than half a billion people and mapping the social relations among them, for creating a new system of exchanging information and for changing how we live our lives. *Time* summarized Facebook's mission: "Facebook wants to populate the wilderness, tame the howling mob and turn the lonely, antisocial world of random chance into a friendly world, a serendipitous world. You'll be working and living inside a network of people, and you'll never have to be alone again. The Internet, and the whole world, will feel more like a family, or a college dorm, or an office where your co-workers are also your best friends" [46].

With over one billion active monthly users in 2012, Facebook is in a unique position to influence the world by enabling Facebook users to create grassroots movements for peace. U.S. Navy Admiral James Stavridis, Commander of the U.S. European Command (USEUCOM) and NATO's Supreme Allied Commander Europe (SACEUR), spoke at the TEDGlobal 2012 conference in Edinburgh about reaching out to people through social networks and providing services such as teaching Afghan soldiers to read. Stavridis said, "Instead of building walls to create security, we need to build bridges.... The six largest nations in the world in descending order: China, India, **Facebook**, the United States, Twitter, and Indonesia.... Moving that message [of friendship] is how we connect international, interagency, private, public, and the social net, to help create security.... No one person, no one alliance, no one nation, no one of us is as smart as all of us thinking together" [47].

12.6 Attainable Peace

Every day, Facebook asks thousands of its users in their own language: "Do you think we will achieve world peace within 50 years?" In 2012, only an average of 9.76 % of U.S. respondents believe that world peace is possible [48].

When I was a young teenager, I carried a small notebook with me everywhere I went so that I could jot down new ideas and discoveries of the world around me. The cover of every one of my notebooks had spaces for me to fill in my name and address. I wrote my real name of course but I put down the imaginary address of a city and country named "Peace." I was obsessed with the notion of peace after having read many books on the history of World War II.

Some politicians such as 2012 presidential candidate Mitt Romney gave up on peace at the onset and hoped for a miracle down the road. Romney said, "I look at the Palestinians not wanting to see peace anyway, for political purposes, committed to the destruction and elimination of Israel, and these thorny issues, and I say, 'There's just no way.'... You hope for some degree of stability, but you recognize that this is going to remain an unsolved problem. We live with that in China and Taiwan. All right, we have a potentially volatile situation, but we sort of live with it, and we kick the ball down the field and hope that ultimately, somehow, something will happen and resolve it" [49].

Romney's view was in sharp contrast to President John F. Kennedy's speech at the 1963 commencement address at the American University in Washington, D.C. [50]:

> First: Let us examine our attitude toward peace itself. Too many of us think it is impossible. Too many think it unreal. But that is a dangerous, defeatist belief. It leads to the conclusion that war is inevitable—that mankind is doomed—that we are gripped by forces we cannot control. We need not accept that view. Our problems are manmade—therefore they can be solved by man. And man can be as big as be wants. No problem of human destiny is beyond human beings. Man's reason and spirit have often solved the seemingly unsolvable, and we believe they can do it again.

Martin Luther King Jr. echoed Kennedy's belief in his 1964 Nobel Peace Prize acceptance speech in which he said, "I refuse to accept the view that mankind is so tragically bound to the starless midnight of racism and war that the bright daybreak of peace and brotherhood can never become a reality" [51].

There are recent examples confirming the optimism of Kennedy and King that peace can prevail: In response to the September 2012 protests across the Arab countries against the anti-Islamic film *Innocence of Muslims*, the Grand Mufti of Egypt Ali Gomaa said, "My message to those who want [strife] between Muslims and Christians in Egypt, I tell them, 'You will not succeed, because we are one people that have been living together for more than 1,400 years'" [52]. In October 2012, Philippines President Benigno Aquino announced a historic peace deal after 15 years of negotiations between the government and the Moro Islamic Liberation Front. Aquino said, "This means that hands that once held rifles will be put to use tilling land, selling produce, manning work stations and opening doorways of opportunity for other citizens" [53].

12.7 A Just and Lasting Peace

At the 1963 American University commencement address, President John F. Kennedy said that peace is "not a Pax Americana enforced on the world by American weapons of war" [50]. Instead, peace is the result of international cooperation and mutual tolerance:

> Let us focus instead on a more practical, more attainable peace—based not on a sudden revolution in human nature but on a gradual evolution in human institutions—on a series of concrete actions and effective agreements which are in the interest of all concerned. There is no single, simple key to this peace–no grand or magic formula to be adopted by one or two powers. Genuine peace must be the product of many nations, the sum of many acts. It must be dynamic, not static, changing to meet the challenge of each new generation. For peace is a process—a way of solving problems.
> With such a peace, there will still be quarrels and conflicting interests, as there are within families and nations. World peace, like community peace, does not require that each man love his neighbor—it requires only that they live together in mutual tolerance, submitting their disputes to a just and peaceful settlement. And history teaches us that enmities between nations, as between individuals, do not last forever.

In his 2009 Nobel Peace Prize acceptance speech, President Barack Obama outlined three ways to build a "just and lasting" peace [9]:

First, in dealing with those nations that break rules and laws, I believe that we must develop alternatives to violence that are tough enough to actually change behavior—for if we want a lasting peace, then the words of the international community must mean something. Those regimes that break the rules must be held accountable. Sanctions must exact a real price. Intransigence must be met with increased pressure—and such pressure exists only when the world stands together as one.... The closer we stand together, the less likely we will be faced with the choice between armed intervention and complicity in oppression.

This brings me to a second point—the nature of the peace that we seek. For peace is not merely the absence of visible conflict. Only a just peace based on the inherent rights and dignity of every individual can truly be lasting. It was this insight that drove drafters of the Universal Declaration of Human Rights after the Second World War. In the wake of devastation, they recognized that if human rights are not protected, peace is a hollow promise.... I believe that peace is unstable where citizens are denied the right to speak freely or worship as they please; choose their own leaders or assemble without fear.... Only when Europe became free did it finally find peace.

Third, a just peace includes not only civil and political rights—it must encompass economic security and opportunity. For true peace is not just freedom from fear, but freedom from want. It is undoubtedly true that development rarely takes root without security; it is also true that security does not exist where human beings do not have access to enough food, or clean water, or the medicine and shelter they need to survive. It does not exist where children can't aspire to a decent education or a job that supports a family. The absence of hope can rot a society from within. And that's why helping farmers feed their own people—or nations educate their children and care for the sick—is not mere charity.

Obama concluded his speech by saying, "As the world grows smaller, you might think it would be easier for human beings to recognize how similar we are; to understand that we're all basically seeking the same things; that we all hope for the chance to live out our lives with some measure of happiness and fulfillment for ourselves and our families... The non-violence practiced by men like Gandhi and King may not have been practical or possible in every circumstance, but the love that they preached—their fundamental faith in human progress—that must always be the North Star that guides us on our journey. For if we lose that faith—if we dismiss it as silly or naïve; if we divorce it from the decisions that we make on issues of war and peace—then we lose what's best about humanity.... Clear-eyed, we can understand that there will be war, and still strive for peace. We can do that—for that is the story of human progress; that's the hope of all the world; and at this moment of challenge, that must be our work here on Earth" [9].

The enormous amount of financial resources and creative energy that nations have spent on wars and weapons could have been redirected to curing deadly diseases, feeding the hungry, eliminating poverty, promoting art and culture, investing in renewable clean energy, and solving a host of other important challenges facing humanity.

President John F. Kennedy said in his 1961 inaugural address, "So let us begin anew —remembering on both sides that civility is not a sign of weakness, and sincerity is always subject to proof.... Let both sides explore what problems unite us instead of belaboring those problems which divide us" [29].

On July 20, 1969, Apollo 11 landed on the moon when an estimated 600 million people around the world were watching on live television. Astronaut Neil Armstrong climbed down the *Eagle*'s ladder and proclaimed: "That's one small step for a man, one giant leap for mankind" [54]. On February 11, 2013, Army Staff Sgt. Clint Romesha became the fourth living person to receive the Medal of Honor for his uncommon valor in Afghanistan [55]. Romesha told reporters after the White House ceremony, "I stand here with mixed emotions of both joy and sadness today. The joy comes from recognition from us doing our jobs as soldiers on distant battlefields. But it is countered by the constant reminder of the loss of our battle buddies—my battle buddies, my soldiers, my friends" [56]. Romesha declined First Lady Michelle Obama's invitation to be her guest at the 2013 State of the Union Address. Instead he spent the evening with families and friends from his former unit, Black Knight Troop, 3–61 CAV [57].

If we take a small step in extolling peacemakers as much as honoring war heroes, we will be making a giant leap towards peace. There are true stories about how soldiers risk their lives for the enemy in acts of military chivalry through the ages. Prof. Shannon E. French at Case Western Reserve University and the U.S. Naval Academy spoke about the warrior's code and cultures: "There is something worse than death, and one of those things is to completely lose your humanity" [58, 59].

Astronauts Neil Armstrong, Buzz Aldrin, and Michael Collins walked on the moon and left behind a plaque that reads, "Here men from the planet Earth first set foot upon the moon. July 1969 A.D. We came in peace for all mankind" [60]. Let us all reach for the higher and more positive side of the human equation!

References

1. **FBI.** What We Investigate. [Online] Federal Bureau of Investigation. [Cited: December 28, 2012.] http://www.fbi.gov/albuquerque/about-us/what-we-investigate.
2. **Chomsky, Noam.** The United States is a Leading Terrorist State. [Online] Monthly Review, November 2001. http://monthlyreview.org/2001/11/01/the-united-states-is-a-leading-terrorist-state.
3. **Maher, Bill.** Spacial Delivery. [Online] HBO, November 30, 2012. http://www.real-time-with-bill-maher-blog.com/real-time-with-bill-maher-blog/2012/11/30/spacial-delivery.html.
4. **CBS Interactive.** World War II Casualties. [Online] CBS News, May 6, 2005. http://www.cbsnews.com/elements/2005/05/06/in_depth_world/frameset693544.shtml.
5. **The Telegraph.** The Holocaust death toll. [Online] The Telegraph, January 26, 2005. http://www.telegraph.co.uk/news/1481975/The-Holocaust-death-toll.html.
6. **Yamazaki, James N.** Hiroshima and Nagasaki Death Toll. [Online] University of California Los Angeles, October 10, 2007. http://www.aasc.ucla.edu/cab/200708230009.html.
7. **Dear, John S.J.** Afghanistan journal, part two: bearing witness to peacemaking in a war-torn country. [Online] National Catholic Reporter, December 18, 2012. http://ncronline.org/blogs/road-peace/afghanistan-journal-part-two-bearing-witness-peacemaking-war-torn-country.
8. **Mount, Mike.** U.S. may remove all troops from Afghanistan after 2014. [Online] CNN, January 8, 2013. http://security.blogs.cnn.com/2013/01/08/u-s-may-remove-all-troops-from-afghanistan-after-2014/.
9. **Obama, Barack.** Remarks by the President at the Acceptance of the Nobel Peace Prize. [Online] The White House, December 10, 2009. http://www.whitehouse.gov/the-press-office/remarks-president-acceptance-nobel-peace-prize.

10. **Davis, William C.** The Battlefields of the Civil War: The Bloody Conflict of North Against South Told Through the Stories of Its Battles. [Online] University of Oklahoma Press, 1991. http://books.google.com/books?isbn=0806128828.

11. **Bronner, Ethan and Kirkpatrick, David D.** U.S. Seeks Truce on Gaza as Enemies Step Up Attacks. [Online] The New York Times, November 20, 2012. http://www.nytimes.com/2012/11/21/world/middleeast/israel-gaza-conflict.html?pagewanted=all.

12. **Newton, Paula.** Analysis: Conflict shifts balance of power in the Middle East. [Online] CNN, November 22, 2012. http://edition.cnn.com/2012/11/21/world/meast/middle-east-balance-power/index.html.

13. **Greenwood, Phoebe.** Gaza declares 'victory' as ceasefire holds, but the most evident triumph is one of survival . [Online] The Telegraph, November 22, 2012. http://www.telegraph.co.uk/news/worldnews/middleeast/palestinianauthority/9697326/Gaza-declares-victory-as-ceasefire-holds-but-the-most-evident-triumph-is-one-of-survival.html.

14. **King, Martin Luther Jr.** The Nobel Peace Prize 1964: Martin Luther King Jr.: Nobel Lecture. [Online] The Nobel Foundation, December 11, 1964. http://www.nobelprize.org/nobel_prizes/peace/laureates/1964/king-lecture.html.

15. **D'Agata, Charlie.** Civilians caught in Israel-Gaza conflict. [Online] CBS News, November 18, 2012.http://www.cbsnews.com/8301-18563_162-57551583/civilians-caught-in-israel-gaza-conflict/.

16. **CNN Wire Staff.** Cease-fire appears to be holding in Gaza. [Online] CNN, November 22, 2012. http://www.cnn.com/2012/11/21/world/meast/gaza-israel-strike/index.html.

17. **ICRC Resource Center.** What is international humanitarian law? [Online] International Committee of the Red Cross, July 31, 2004. http://www.icrc.org/eng/resources/documents/legal-fact-sheet/humanitarian-law-factsheet.htm.

18. **Urquhart, Conal.** The Goldstone report: a history. [Online] The Guardian, April 14, 2011. http://www.guardian.co.uk/world/2011/apr/14/goldstone-report-history.

19. **Rudoren, Jodi and Kershner, Isabel.** Israel Broadens Its Bombing in Gaza to Include Government Sites. [Online] The New York Times, November 17, 2012. http://www.nytimes.com/2012/11/18/world/middleeast/israel-gaza-conflict.html?pagewanted=all.

20. **Sanger, David E. and Shanker, Thom.** For Israel, Gaza Conflict Is Test for an Iran Confrontation. [Online] The New York Times, November 22, 2012. http://www.nytimes.com/2012/11/23/world/middleeast/for-israel-gaza-conflict-a-practice-run-for-a-possible-iran-confrontation.html?pagewanted=all.

21. **Tolstoy, Leo.** War and Peace. [Online] 1869. http://books.google.com/books?id=jhZzwKsi0OsC.

22. **IMDb.** Sherlock Holmes: A Game of Shadows. [Online] IMDb, December 16, 2011. http://www.imdb.com/title/tt1515091/.

23. **Sidner, Sara.** Children of the conflict: Innocence interrupted by war. [Online] CNN, November 26, 2012. http://www.cnn.com/2012/11/25/world/meast/sidner-children-israel-gaza-conflict/index.html.

24. **U.S. Secret Service.** Protective Mission. [Online] United States Secret Service. [Cited: December 29, 2012.] http://www.secretservice.gov/protection.shtml.

25. **U.S. Department of Defense.** Army Releases November Suicide Data. [Online] U.S. Department of Defense, December 13, 2012. http://www.defense.gov/releases/release.aspx?releaseid=15741.

26. **Watkins, Tom and Schneider, Maggie.** 325 Army suicides in 2012 a record. [Online] CNN, February 2, 2013. http://www.cnn.com/2013/02/02/us/army-suicides/index.html.

27. **U.S. Department of Defense.** Army Releases July Suicide Data. [Online] U.S. Department of Defense, August 16, 2012. http://www.defense.gov/releases/release.aspx?releaseid=15517.

28. **TED.** The technological future of crime: Marc Goodman at TEDGlobal 2012. [Online] TED, June 28, 2012. http://blog.ted.com/2012/06/28/the-technological-future-of-crime-marc-goodman-at-tedglobal-2012/.

29. **Kennedy, John F.** Ask not what your country can do for you. [Online] The Guardian, January 20, 1961. http://www.guardian.co.uk/theguardian/2007/apr/22/greatspeeches.

30. **The Constitutional Convention .** Constitution of the United States. [Online] U.S. National Archives and Records Administration, September 17, 1787. http://www.wdl.org/en/item/2708/.

31. **BBC.** Malala Yousafzai: Portrait of the girl blogger . [Online] BBC News Magazine, October 10, 2012. http://www.bbc.co.uk/news/magazine-19899540.

32. **Leiby, Richard and Leiby, Michele Langevine.** Taliban says it shot Pakistani teen for advocating girls' rights. [Online] The Washington Post, October 9, 2012. http://www.washingtonpost.com/world/asia_pacific/taliban-says-it-shot-infi- del-pakistani-teen-for-advocating-girls-rights/2012/10/09/29715632-1214- 11e2-9a39-1f5a7f6fe945_story.html.

33. **Hays, Julie.** Pakistani teen inspires others to fight for education. [Online] CNN, October 15, 2012. http://www.cnn.com/2012/10/15/world/iyw-support-for-malala/index.html.

34. **Bush, Laura.** A girl's courage challenges us to act. [Online] The Washington Post, October 10, 2012. http://www.washingtonpost.com/opinions/laura-bush- malala-yousafzais-courage-challenges-us-to-act/2012/10/10/9cd423ea-1316-11e2 -ba83-a7a396e6b2a7_story.html.

35. **Kimball, Roger.** The Groves of Ignorance. [Online] 1987, 5 April. http://www.nytimes. com/1987/04/05/books/the-groves-of-ignorance.html?pagewanted=all.

36. **Chomsky, Noam and Schivone, Gabriel Matthew.** United States of Insecurity: Interview with Noam Chomsky. [Online] Monthly Review, May 2008. http://monthlyreview. org/2008/05/01/united-states-of-insecurity-interview-with-noam-chomsky.

37. **Facebook.** Peace on Facebook. [Online] Facebook. [Cited: January 1, 2013.] https:// peace.facebook.com/.

38. **Rainie, Lee and Smith, Aaron.** Social networking sites and politics . [Online] Pew Internet, March 12, 2012. http://www.pewinternet.org/Reports/2012/Social-networking-and-politics/ Main-findings/Social-networking-sites-and-politics.aspx.

39. **CNN Editors.** Syria's 'cyber warriors' choose cameras over guns. [Online] CNN, June 14, 2012. http://globalpublicsquare.blogs.cnn.com/2012/06/14/syrias-cyber-warriors-choose-cam eras-over-guns/.

40. **—.** Syria's 'cyber warriors' choose cameras over guns. [Online] CNN, June 14, 2012. http://- globalpublicsquare.blogs.cnn.com/2012/06/14/syrias-cyber-warriors-choose-cameras-over-g uns/.

41. **Newton-Small, Jay.** Hillary's Little Startup: How the U.S. Is Using Technology to Aid Syria's Rebels. [Online] Time Magazine, June 13, 2012. http://world.time.com/2012/06/13/hillarys- little-startup-how-the-u-s-is-using-technology-to-aid-syrias-rebels/.

42. **The American-Israeli Cooperative Enterprise.** The Arab-Israeli Conflict: Total Casualties (1920-2012). [Online] Jewish Virtual Library. [Cited: April 20, 2012.] http://www.jewishvirtu allibrary.org/jsource/History/casualtiestotal.html.

43. **Strategic Foresight Group.** Cost of Conflict in the Middle East. [Online] Strategic Foresight Group Report Excerpts, January 2009. http://www.strategicforesight.com/Cost%20of%20 Conflict%20-%206%20pager.pdf.

44. **Mehina, Pushpin.** Pushpin Mehina. [Online] Facebook. [Cited: April 20, 2012.] http://www. facebook.com/pushpin.

45. **Said, Samira.** Peace-minded Israeli reaches out to everyday Iranians via Facebook. [Online] CNN, March 20, 2012. http://www.cnn.com/2012/03/19/world/meast/israel-iran-social-media/ index.html.

46. **Grossman, Lev.** Person of the Year 2010. Mark Zuckerberg. [Online] Time Magazine, December 15, 2010. http://www.time.com/time/specials/packages/article/0,28804,2036683_2 037183_2037185,00.html.

47. **Stavridis, James.** James Stavridis: A Navy Admiral's thoughts on global security. [Online] TED, June 2012. http://www.ted.com/talks/james_stavridis_how_nato_s_supreme_commander_thinks_ about_global_security.html.

48. **Facebook.** Peace on Facebook. [Online] Facebook. [Cited: January 1, 2013.] https:// peace.facebook.com/.

49. **Wilson, Scott and O'Keefe, Ed.** Mitt Romney: 'Palestinians have no interest whatsoever in establishing peace'. [Online] The Washington Post, September 18, 2012. http://articles.washi ngtonpost.com/2012-09-18/politics/35497236_1_peace-talks-mitt-romney-palestinians.

50. **Kennedy, John F.** Commencement Address at American University, June 10, 1963. [Online] John F. Kennedy Presidential Library and Museum, June 10, 1963. http://www.jfklibrary.org/Researchold/Ready-Reference/JFK-Speeches/Commencement-Address-at-American-University-June-10-1963.aspx.
51. **King, Martin Luther Jr.** Martin Luther King's Acceptance Speech. [Online] The Nobel Foundation, 1964. http://www.nobelprize.org/nobel_prizes/peace/laureates/1964/king-acceptance_en.html.
52. **Lee, Ian and Fahmy, Mohamed Fadel.** Sunni Islam leader calls for peace, urges Muslims to have 'patience and wisdom' . [Online] CNN, November 19, 2012. http://www.cnn.com/2012/09/22/world/world-film-protests/index.html.
53. **CNN Wire Staff.** Philippines, Muslim rebels reach peace deal . [Online] CNN, October 7, 2012. http://www.cnn.com/2012/10/07/world/asia/philippines-peace-deal/index.html.
54. **The Telegraph.** Apollo 11 Moon landing: ten facts about Armstrong, Aldrin and Collins' mission. [Online] The Telegraph, July 18, 2009. http://www.telegraph.co.uk/science/space/5852237/Apollo-11-Moon-landing-ten-facts-about-Armstrong-Aldrin-and-Collins-mission.html.
55. **Tapper, Jake and Carter, Chelsea J.** An American hero: The uncommon valor of Clint Romesha. [Online] CNN, February 8, 2013. http://www.cnn.com/2013/02/11/politics/medal-of-honor/index.html.
56. **Tapper, Jake and Cohen, Tom.** Medal of Honor recipient conflicted by joy, sadness. [Online] CNN, February 12, 2013. http://www.cnn.com/2013/02/11/politics/medal-of-honor/index.html.
57. **Tapper, Jake.** Medal of Honor recipient declines invitation to State of the Union. [Online] CNN, February 12, 2013. http://www.cnn.com/2013/02/12/politics/sotu-invite-declined/index.html.
58. **Blake, John.** Two enemies discover a 'higher call' in battle. [Online] CNN, March 9, 2013. http://www.cnn.com/2013/03/09/living/higher-call-military-chivalry/index.html
59. **French, Shannon E.** The Warrior's Code. U.S. Naval Academy. [Online] International Society for Military Ethics, 2001. http://isme.tamu.edu/JSCOPE02/French02.html
60. **NASA.** July 20, 1969: One Giant Leap For Mankind. [Online] National Aeronautics and Space Administration, July 8, 2009. http://www.nasa.gov/mission_pages/apollo/apollo11_40th.html.

About the Author

Newton Lee is CEO of Newton Lee Laboratories LLC, president of the Institute for Education, Research, and Scholarships, and founding editor-in-chief of ACM Computers in Entertainment. Previously, he was a research scientist at Bell Laboratories, senior producer and engineer at The Walt Disney Company, and research staff member at the Institute for Defense Analyses where he co-pioneered artificial intelligence applications in counterterrorism. Lee graduated Summa Cum Laude from Virginia Tech with a B.S. and M.S. degree in Computer Science, and he earned a perfect GPA from Vincennes University with an A.S. degree in Electrical Engineering and an honorary doctorate in Computer Science. He is the co-author of *Disney Stories: Getting to Digital* and the author of the Total Information Awareness book series including *Facebook Nation* and *Counterterrorism and Cybersecurity*.

N. Lee, *Counterterrorism and Cybersecurity*, DOI: 10.1007/978-1-4614-7205-6,
© Springer Science+Business Media New York 2013

Index

Printed in the United States
By Bookmasters